OBSESSIVE-COMPULSIVE DISORDER - THE OLD AND THE NEW PROBLEMS

Edited by **Vladimir Kalinin**

Obsessive-Compulsive Disorder - The Old and the New Problems
http://dx.doi.org/10.5772/56999
Edited by Vladimir Kalinin

Contributors

Gaikwad Uday, María-José Martín-Vázquez, Zena Al-Sharbati, Elisa Krackow, Mushtaq Margoob, Marc Lavoie, Geneviève Sauvé, Simon Morand-Beaulieu, Kieron O'Connor, Elana Harris, Milena Korostenskaja, Vladimir Kalinin

Published by InTech

Janeza Trdine 9, 51000 Rijeka, Croatia

Publishing Process Manager Iva Lipovic

Technical Editor InTech DTP team

Cover InTech Design team

First published March, 2014

Printed in Croatia

Additional hard copies can be obtained from orders@intechopen.com

Obsessive-Compulsive Disorder - The Old and the New Problems, Edited by Vladimir Kalinin

p. cm.

ISBN 978-953-51-1238-9

For EU product safety concerns:
IN TECH d.o.o., Prolaz Marije Krucifikse Kozulić 3, 51000 Rijeka, Croatia,
info@intechopen.com or visit our website at intechopen.com.

Meet the editor

Vladimir Kalinin, MD, PhD, was born in 1952 in Orenburg, Russian Federation, in family of physicians. He graduated from Moscow Medical Stomatological University in 1976 and in 1976-1977 he finished a psychiatry internship. In 1978 he became a scientific researcher at Moscow Research Institute of Psychiatry, where he is still working as Head of Department. The scope of his scientific interests concerns a broad range of psychiatry problems, including general psychopathology, psychopharmacology, neuropsychiatry and epileptology. He is author of 180 publications, including research articles in professional journals and 4 monographs. He is also the editor of the book "Anxiety disorders" published by InTech in 2011. He is a member of Presidium of Russian Psychiatric Society, and member of Russian League against Epilepsy.

Contents

Preface

Obsessive-compulsive disorder (OCD) represents really unique category in psychopathology since it is characterized by signs that are similar, but not identical to other psychopathological syndromes. Thus, OCD is usually accompanied by unpleasant affects such as sadness, anxiety, and fear that cause the patients the feelings of suffering and are *Ego-dystonic*. On the other hand, the structure of obsessions itself is rather similar to delusions and so-called thought automatisms, but unlike the latter, the insight in OCD patients is preserved, and patients themselves really understand the morbid nature of their suffering.

In his excellent book on the history of psychopathological symptoms, a famous modern psychiatrist German Berrios (2002) regards the long way of formation of terms "Obsessions and compulsions" in their modern sense. Thus, he points out that during the 19th century, obsessions were considered as signs of affective or, alternatively, as pathology of thinking. Obviously, such inconsistency in appraisal of obsessions may be explained by overlap between affective, delusional symptoms and obsessions themselves.

Noteworthy, OCD shares with depression and anxiety not only common psychopathological features, but neurochemical mechanism too. From historical point of view, the observation of Fernando-Cordoba and Lopez-Ibor Alino (1967) about Clomipramine efficacy in the treatment of OCD patients was the first one. This finding has triggered a great number of studies of the drug treatment of OCD, with emphasis on selective serotonin reuptake inhibitors (SSRI) use. Obtained results have shown the efficacy of SSRI but, unfortunately, not in all cases, and a certain number of OCD cases remain resistant to SSRI treatment.

Principally, response rate on SSRI in OCD patients achieves only 50-80%, and total obsession and compulsion score can't be reduced more than 40-50% compared with baseline level (Hohagen, 2000). In this context, other biochemical mechanisms of OCD origin have been suggested. It concerns the glutamate metabolism dysregulation in terms of increased glutamatergic activity in CNS (Pittenger C., Bloch M.H., Williams K., 2011). The probable efficacy of glutamate-modulating agents in the OCD treatment should be proved in blind placebo-controlled studies.

Many years ago a famous Russian psychiatrist Mark Y. Sereysky (1958) considered OCD syndrome in patients with schizophrenia as buffer one. By this he implied both poor response of OCD syndrome to treatment and lack of any further progression and evolvement to delusional and hallucinatory syndromes. The mechanisms of such properties remain unknown until present days. Nevertheless, such buffer mechanisms can explain the definite and peculiar psychopathological structure, stability and persistence of OCD symptoms.

It should be stressed that pathogenic mechanisms relating OCD with serotonin dysregulation in CNS couldn't explain the unique OCD symptoms which can be simply distinguished from symptoms of affective and anxiety syndromes on psychopathological level. Obviously, further investigations in this context are needed in order to shed light on the OCD pathogenesis. Perhaps, the data connecting OCD pathogenesis with Cortico-Basal Ganglia Circuit Dysfunction would properly explain the OCD mechanisms in the nearest future (Graybiel A.M., Rauch S.L., 2000).

The existing data on cognitive dysfunction in OCD patients, including tests on memory, planning, decision making, executive control, attention and vigilance as a whole confirm the above mentioned tests for the role of endophenotypic markers of OCD (Chamberlain S.R., Blackwell A.D., Fineberg N.A. et al., 2005).

Moreover, the so-called psychotherapeutic approach to the OCD treatment has been elaborated several decades ago, and promising results in this domain have been achieved. Such approach really can broaden our therapeutic success in the OCD treatment.

OCD cause a serious distress, and patients suffering from it usually lose their social adjustment. In numerous studies it has been shown that patients with OCD experience significant disruption in social and family relationships, impaired work performance, and decreased quality of life (Turner S.M. et al., 2004). In this context the role of social support and cognitive behavioral therapy can't be exaggerated and such approach must be widely used for OCD patients.

The present collective monograph contains several chapters concerning the different dimensions of OCD, i.e. unresolved problems of pathogenic mechanisms, origin, treatment and social support work. The principally novel data with emphasis on broad range of yet unresolved OCD problems are presented. The book is intended for medical specialists, mainly for psychiatrists and psychotherapists. In addition, the book may also interest psychologists and social workers. Besides, the data on pathogenesis may be of practical use for physiologists, biochemists, pharmacologists.

References

1. Berrios G.E. The history of mental symptoms. Descriptive psychopathology since the nineteenth century. Cambridge University Press, 2002 – 565 pp.

2. Chamberlain S.R., Blackwell A.D., Fineberg N.A. , Robbins T.W.,Sahakian B.J. The neuropsychology of obsessive compulsive disorder: the importance of failures in cognitive and behavioural inhibition as candidate endophenotypic markers. Neuroscience & biobehavioral reviews, 2005, 29: 399 - 419.

3. Fernando-Cordoba E., Lopez-Ibor Alino J. La monoclomipramina en enfermos psiquitricos resistentes a otros tratamientos. Acta Luso-Esp. Neurol. Psiquiat. Ciene Afines, 1967,26: 119-147.

4. Graybiel A.M., Rauch S.L. Towards a neurobiology of obsessive-compulsive disorder. Neuron, 2000, 28: 343 -347.

5. Hohagen F. Zwangsstörung. In: Förstl H.(Hrsg): Klinische Neuro-Psychiatrie. Neurologie psychischer Störungen und Psychiatrie neurologischer Erkrankrungen. Georg Thieme Verlag, Stuttgart, 2000: 47 – 60. (In Germain).

6. Pittenger C., Bloch M.H., Williams K. Glutamate abnormalities in obsessive compulsive disorder: Neurobiology, pathophysiology, and treatment. Pharmacology & Therapeutics, 2011, 232: 314 – 332.

7. Sereysky M.Y. An obsessional states in schizophrenia. In: Banshikov V.M. (Ed) The problems of clinic, pathogenesis and treatment of schizophrenia, Moscow, 1958: 205 – 212, (In Russian).

8. Turner S. M., Beidel D.C., Stanley M.A., Heiser N. Obsessive-compulsive disorder. In: Adams H.E., Sutker P.B. (Eds) Comprehensive handbook of psychopathology. Springer, New York, 2004: 155 – 182.

Professor Vladimir V. Kalinin
Head of Department of Brain Organic Disorders and Epilepsy,
Moscow Research Institute of Psychiatry of Ministry of Healthcare,
Russian Federation

Section 1

Phenomenology and Pathophysiology of Obsessive-Compulsive Disorder

Chapter 1

Pathophysiology of Obsessive–Compulsive Disorder: Affected Brain Regions and Challenge Towards Discovery of Novel Drug Treatment

Uday Gaikwad

Additional information is available at the end of the chapter

http://dx.doi.org/10.5772/57193

1. Introduction

Obsessive–compulsive disorder (OCD) is a mental disorder characterized by absurd, recurrent and uncontrollable thoughts (obsessions) that produce anxiety, which are followed by repetitive behaviors (compulsions) aimed at reducing anxiety. OCD may be looked upon as a condition in which the affected person frequently experiences irresistible urges to perform repetitive rituals (compulsions). OCD may be defined as the irruption in the mind of uncontrollable, egodystonic and recurrent thoughts, impulses or images. In OCD, repetitive rituals serve to counteract the anxiety precipitated by obsessions. The OCD patients realize the irrational nature of thoughts and rituals but feel helpless and hopeless about controlling them. Obsessive-Compulsive disorder can impair all areas of brain functioning and produce devastating effects on patients and their families. Classic psychoanalysis, as pioneered by Freud, interpreted obsessive-compulsive disorder as unconscious conflicts, which were defensive and punitive (Rapoport et al., 1993). In modern psychoanalysis, obsessive-compulsive disorder is described as a portrayal of ambivalence, with confusion of thoughts and actions that are paradoxically manifested by rigidity and abnormal behaviors. Dynamic psychiatry interprets obsessive-compulsive symptoms as a reflection of feelings and thoughts that provoke aggressive or sexual actions that might produce shame, weakness, or loss of pride (Baer, 1993). The thoughts and behaviors associated with OCD are viewed as senseless, and egodystonic and they stand contradictory to the individual's motives, goals, identity, and self-perception thereby creating significant subjective distress. The excessive nature of the compulsion, however, creates its own distress and it appears that the individual may be caught up in a kind of negative reinforcement loop (David et al., 2004) The obsessive-compulsive spectrum disorders are Tourette's disorder, Body dysmorphic disorder, Hypochondriasis,

Pathological jealousy, Trichotillomania, Skin picking, Nail biting, Compulsive buying, Kleptomania, Pathological gambling, Nonparaphilic sexual disorders, Obsessive compulsive personality disorder.

OCD is a psychiatric affliction with a lifetime prevalence of 1–3% (Rasmussen and Eisen, 1992; Sasson et al., 1997). According to the Diagnostic and Statistical Manual of Mental Disorders (4th ed; DSM IV), the essential features of OCD are recurrent obsessions and/or compulsions (e.g., doubting, checking, washing) that are time consuming (i.e., they take more than 1 h a day) or cause marked distress or significant impairment. To date, the most effective treatments for OCD are pharmacological treatment, using serotonin reuptake inhibitors (SRIs, e.g., Masand and Gupta, 1999; Piccinelli et al., 1995; Pigott and Seay, 1999; Stein et al., 1995; Zohar et al., 1992), and behavioral treatment, using the response exposure and prevention technique (e.g., Simpson et al., 2004). Yet, around 30% of the patients are refractory to pharmaco- and behavioral therapy (Eddy et al., 2004). Some of these treatment-resistant patients are treated by lesions to structures and pathways within basal gangliathalamo- cortical circuits (Lopes et al., 2004) as well as by high frequency stimulation (HFS) of the ventral striatum region (Aouizerate et al., 2004, 2005; Greenberg et al., 2006, 2008; Rauch et al., 2006; Sturm et al., 2003) the subthalamic nucleus (Mallet et al., 2008), and the thalamic reticular nucleus and the inferior thalamic peduncles (Jimenez et al., 2007; Jimenez-Ponce et al., 2009). Several neural systems have been implicated in the pathophysiology of OCD: The results of neuroimaging studies in OCD patients have implicated most consistently the orbitofrontal cortex, the cingulated cortex and the basal ganglia, and more recently also regions within the parietal lobe, in the pathophysiology of obsessions and compulsions (for review see Menzies et al., 2008; Rotge et al., 2009; Saxena et al., 1998; Stein, 2000). Dysregulation of the serotonergic (5-HT) system has been suggested primarily on the basis of the effectiveness of SRI's and selective serotonin reuptake inhibitors (SSRI's) in alleviating obsessions and compulsions in patients (Zohar and Insel, 1987; Zohar et al., 1992), and has received further support from neurobiological, pharmacological and more recently genetic data (for review see Murphy et al., 2001; Ozaki et al., 2003; Sasson and Zohar, 1996; Stein, 2000, but see Baumgarten and Grozdanovic, 1998). Abnormalities of the dopaminergic system have also been implicated in the pathophysiology of OCD, based on surplus therapeutic benefits obtained with co-administration of SSRI's and dopamine blockers (McDougle et al., 1990, 1994; Sasson andZohar, 1996) as well as on clinical observations of obsessions and compulsions in basal ganglia-related disorders, such as Tourette's syndrome (Frankel et al., 1986; Grad et al., 1987; Pitman et al., 1987). More recently, an increasing body of evidence points also to the involvement of the glutamatergic system in OCD (for review, see Pittenger et al., 2006), including association of certain polymorphisms in the NMDA receptor gene with susceptibility to OCD N. Albelda, D. Joel / Neuroscience and Biobehavioral Reviews 36 (2012) 47–63 49 (Arnold et al., 2004); elevated glutamate levels in the cerebro-spinal fluid of drug-naïve patients (Chakrabarty et al., 2005); correlations between symptom severity and the level of several glutamatergic metabolites (Starck et al., 2008); improvement of symptoms following treatment with d-cycloserine (DCS), a partial NMDA agonist (blinded controlled trials, Kushner et al., 2007; Wilhelm et al., 2008), riluzole, a glutamatergic antagonist (open-label trials, Coric et al., 2005; Grant et al., 2007), and memantine, a non-competitive NMDA antagonist (an open-label trial, Aboujaoude et al., 2009). There is also some evidence suggesting the involvement of nitric oxide (NO) in OCD. Atmaca et al. (2005) found that OCD patients have higher NO levels in their plasma compared to healthy subjects and that these levels are positively correlated with the severity of OC symptoms. The possibility that high NO levels are related to OC symptoms is supported by the fact that SSRI's, anti-dopaminergic drugs and the NMDA antagonist memantine, all used to treat OCD patients, inhibit the synthesis of NO (Almeida et al., 2006; Park and West, 2009; Zhang et al., 2010). Reports that life events related to the female hormonal cycle may trigger or exacerbate OCD in women patients (Abramowitz et al., 2003; Labad et al., 2005; Maina et al., 1999) suggest that ovarian hormones play a modulatory role in OCD (Uguz et al., 2007). Indeed, gonado-tropine-releasing hormone (GnRH) agonists were reported to ameliorate OC symptoms in OCD patients (Casas et al., 1986; Eriksson, 2000). The understanding and treatment of diseases such as OCD must rely heavily on appropriate animal models that closely mimic their behavioral and if possible their neural manifestations. This is especially true for OCD as its neuropathological mechanisms are still largely unknown, and many patients are either treatment-resistant or experience only partial alleviation of symptoms. Before reviewing animal models of OCD that are currently in use, we discuss the criteria for the validation and evaluation of animal models.

2. Clinical features

The OCD is clinically manifested by a wide range of symptoms. The most common types of obsessions are related to contamination, pathological doubts, somatic dysfunctions, need for symmetry, aggression and hyper sexual drive. The classical forms of compulsions include checking, washing, counting, need to ask, precision and hoarding. In OCD, senseless, repetitive rituals (such as counting, washing etc.) serve to counteract the anxiety precipitated by obsessive thoughts e.g. Symmetry and exactness preoccupations. Fears of contamination and illness produce washing and cleaning compulsions (Leckman et al., 1997). Some symptoms get stable over a time period i.e. sexual/religious obsessions. The symptoms that is more likely to change are aggressive and contamination obsessions. Changes usually occur within rather than between individual symptom dimensions (Mataix-Cols et al., 2002). Patients with OCD may report only one or, more typically, multiple symptoms that cut across dimensions (Stein et al., 1997).

Many children with obsessive-compulsive disorder (OCD) suffer from an almost pathological doubting, which can vary from a mild form to an incapacitating form of extreme severity, in which the child is uncertain about its own understandings. Other high frequency symptoms are checking, fear of harming others, obsessions relating to death or sex produce compulsions serving to avoid a horrible event. Indecisiveness is frequently found in children with OCD and ranges from difficulty in making minor decisions. Following Group A Beta–Hemolytic Streptococcal (GABHS) infections, it was noted that OCD symptoms along with choreic movements in sydenham chorea became worse, and this was mediated by antineuronal antibodies that adversely affected basal ganglia cells. It was later hypothesized that this abnormal reaction to GABHS might play a part in the etiology of OCD.[1] This subgroup of OCD patients has been termed PANDAS (for Pediatric Autoimmune Neuropsychiatric disorders Associated with Streptococcal Infection). Asbahr, Ramos, Negrao and Gentil presented four cases of children with PANDAS. They suggested that OCD may occur after repeated streptococcal infection as all four cases developed OCD only after the second infection. MRI tests showed that children with PANDAS have enlarged basal ganglia. The frontal-lobe function compared in 21 children and adolescents with OCD who were approximately 12 years of age and in matched healthy subjects, investigators found no cognitive impairment on the basis of performance on frontal tests and concluded that OCD symptoms may not interfere with cognitive abilities at early stages in the illness. The study with 42 children and adolescents (mean age, 14 years) and 35 normal subjects and concluded that subjects with OCD have impaired executive functions and nonverbal memory. Andres administered a neuropsychological battery to 35 OCD patients without psychiatric comorbidity aged between 7 and 18 years and 35 healthy controls and found that children and adolescents with OCD had impairments in visual memory, visual organization.[2]

The performances of 21 obsessive-compulsive children compared with 21 healthy controls on neuropsychological tests, including the Stroop test, Wisconsin Card Sorting Test, Verbal Fluency test, California Verbal Learning Test, Grooved Pegboard Test suggested there were no differences in neuropsychological performance between obsessive-compulsive children and the healthy controls. Therefore, it was concluded that obsessive-compulsive children do not show clinically significant cognitive impairment during the early stages of the illness, but a deficiency may become significant over time. The doubting and checking are found frequently in OCD patients which associated with memory dysfunction of OCD patients. The children with tic disorder frequently show OCD symptoms. The negative emotions such as anxiety have an adverse effect on memory function. The most striking feature of the symptoms presented by obsessive-compulsive children is the severity of the psychopathology in the absence of formal thought disorder. Thus, Obsessive-Compulsive disorder can impair all areas of brain functioning and produce devastating effects on patients and their families. Therefore, there is necessity to aware about the OCD in children and different treatments for it's management.

Sr.	Obsessions	Compulsions
1.	Concern with cleanliness (dirts, germs, contamination)	Excessive and ritualized bathing, washing and cleaning
2.	Concern with exactness (symmetry and order)	Ritualized arranging and rearranging
3.	Concern with household tools (dishes, spoons, soap)	Checking, rechecking and keeping inventory with detailed description of tools, objects, and appliances
4.	Concern about body secretions (saliva, urine, stool)	Rituals to remove body secretions
5.	Sexual obsessions (aggressive sexual actions)	Ritualized and rigid sexual relationship

Table 1. List of typical obsessive thoughts compelling repetitive actions

3. Causes

- Serotonin is involved in regulating anxiety
- Abnormality in the neurotransmitter serotonin
- In order to send chemical messages serotonin must bind to the receptor sites located on the neighboring nerve cells
- OCD suffers may have blocked or damaged receptor sites preventing serotonin from functioning to full potential
- Possible genetic mutation
- Some people suffering have mutation in the human serotonin transporter gene

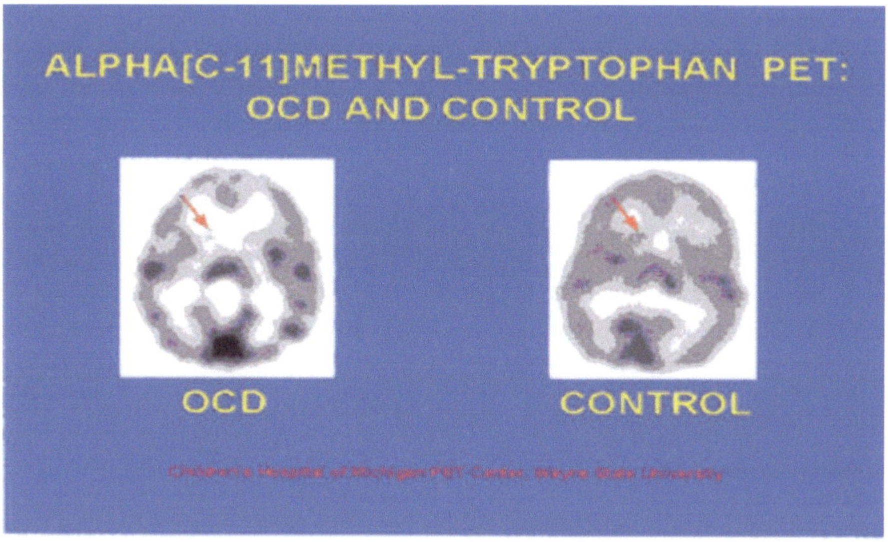

OCD patients have lower serotonin levels than control subjects. Source: http://neuron.med.wayne.edu/OCD/slide.htm

Figure 1. OCD patients have lower serotonin levels than control subjects.

4. Epidemiology

The frequency of compulsions in 8-year-old German children, to be 4.6% (moderate severity) and 2.8% (severe symptomatology). Later, the prevalence of compulsive checking in the same population (13 years old) was 3.4%, all of moderate severity. 2.3% reported compulsion for cleanliness and 4.0% experienced obsessive thoughts. OCD in the child psychiatric population found as 0.2-1.2%. In an epidemiological study of 5000 high-school students, Flament found that 0.35% fulfilled the criteria of OCD, as defined by DSM-III (American Psychiatric Association, 1980). They calculated a lifetime prevalence of 0.4%. In their Isle of Wight-study, 2000 children examined at the age of 11 and found no children who fulfilled the criteria for OCD. They found eight children whom they diagnosed with "mixed emotional disorder" (with obsessive-compulsive features).

5. Genetics and family lifestyle

Family lifestyle could be regarded as a possible precipitating factor, for developing obsessive-compulsive symptoms in predisposed children. Family functioning was described in 20 Danish children with OCD and later assessed adaptability and cohesion of the family. In studies of children and adolescents most surveys show that OCD patients are generally from white, middle- or upper-class, intact families. Certain myths have prevailed regarding families with OCD children. The families are often said to show a cultural behaviour that emphasises cleanliness and perfection. 9% of mothers and 25% of fathers to OCD children had OCD themselves. 19% of 21 OCD children had one parent with OCD and as many as 52% had parents with obsessive-compulsive symptoms, without meeting the criteria for OCD, when parents were interviewed with standard clinical psychiatric assessments. 52% of children with OCD have a positive family history of OCD amongst first-degree relatives and that risk to relatives is related to the age of onset of the OCD proband: the younger the onset of OCD in proband, the higher risk of OCD or Tourette's syndrome in relatives. 24% of referred OCD children had a first degree relative with OCD.

6. Biological aspects

An electrophysiological investigation in obsessional states informed the presence of altered neural inhibition in connection with obsessive symptoms. The EEGs of a group of OCD patients compared with a control group of patients with other neurotic disorders, indicated abnormal findings in two thirds of the OCD patients, which was not significantly higher than in the control group. An orbitofrontal-subcortical hyperactivity in children and adolescent OCD patients may be the result of abnormal neuroanatomical development of these structures. A report suggested OCD develop by head injuries in four cases. The OCD patients have a particularly higher range of neurological diseases in childhood.

7. Neurological soft-signs and neuropsychological findings

Soft-signs are non-localized deviant performances in a motor or sensory test, without other signs or presence of focal neurological disorder. Soft-signs in child psychiatric patients have been well studied and implications of soft-signs have been thouroughly analysed in a general population of pre-school and school-children. 44 out of 54 childhood OCD patients, at the age from 6 to 20 years, had neurological abnormalities. Of these 44 patients 18 had choreiform movements, 13 had non-specific neurodevelopmental signs only, eight had left hemisyndrome, and five had miscellaneous abnormalities. Most studies conducted on OCD children have resulted an overweigth of soft-signs, compared to normal control-groups. The children with early OCD onset, have more soft-signs than children with a later onset. The neurological soft-signs found in 18.6% of 61 OCD-children and adolescents compared to 14.4% in a control group consisting of non-psychotic and non-retarded children and adolescents with psychiatric disorders other than OCD. In studies of children, a pattern of association between OCD and selected neurological disorders has been long recognized. A prominently left frontal dysfunction in OCD patients indicated that cerebral cause had responsible for the obsessive-compulsive symptoms.

8. Pathophysiology

The brain regions impaired in OCD includes dorsolateral prefrontal cortex (DLPC), anterior cingulate cortex (ACC), basal ganglia, orbito-frontal cortex (OFC), striatum, amygdala, thalamus and brainstem.

8.1. Dorsolateral prefrontal cortex (DLPC)

It is the most important cortex part for cognitive functions in human beings. The involvement of the DLPC in working memory was initially demonstrated in primate studies. The DLPC also plays a role in adaptation to changes in the environment. DLPC plays a crucial role in focusing attention on specific stimuli and in decision-making (Miller, 1999). Lesions of DLPC disturb the subject's ability to process temporal information and impair the successful performance of goal-directed behaviors. Functional neuroimaging data have shown diminished activity in the DLPC of patients suffering with psychiatric disorders such as major depression and OCD, which may account for the difficulty in overcoming compulsive behaviors (Saxena et al., 1998).

8.2. Anterior cingulate cortex (ACC)

Neuroimaging studies indicated that the ACC is involved in a variety of cognitive processes such as attention, motivation, reward, error detection, working memory, problem solving and action–plan (Bush et al., 2000). There are two major regions within ACC viz. a dorsal region, known as the cognitive region, and a ventral or affective region. The cognitive region is a part of attentional netwok and is closely connected with the DLPC, premotor, and parietal cortices whereas, the affective region is linked to the amygdala, nucleus accumbens, hypothalamus, anterior insula, hippocampus and OFC and sends projections to the neuro-vegetative, visceromotor and endocrine systems. Excessive activation of ACC has been reported in patients presenting psychiatric disturbances such as phobias, OCD and mood disorders. Moreover, electrophysiological studies in man have demonstrated its particular role in error detection processes.

8.3. Basal ganglia-thalamo-cortical circuits

Basically, the role of the basal ganglia is to integrate the various inputs arriving from the cortex and to use this information for selecting certain motor and/or cognitive programs. The point of entry of information to the basal ganglia is through striatum, which receives converging information from the limbic and associative cortices. It then sends projections to the output structures, i.e. the globus pallidus pars internalis (Gpi) and the substantia nigra pars reticulata (SNr), through two pathways: one direct and the other indirect. The indirect pathway successively involves the globus pallidus pars externalis (Gpe) and the subthalamic nucleus (STN). In addition, the cortex sends direct inputs to the STN and the connections between the Gpe and Gpi.These two pathways seem to play opposite roles in controlling cortical activity. Activation of the direct loop facilitates the triggering of programs at the cortical level through a double inhibition. On the other hand, the indirect loop blocks the activation of thalamic relay by increasing the activity of the Gpi, a GABA-ergic inhibitory structure. Dopamine of nigral origin seems to facilitate the direct pathway through D_1 receptors and plays an inhibitory role on the indirect pathway through D_2 receptors. The pathological activation of segregated closed loop circuits involving cortex-basal ganglia–thalamus–cortex pathways would produce reverberating activity and result in a persistent discharge of the innate programs characteristic of OCD. The clinical manifestations of neuronal disorders of the basal ganglia can be viewed as a disruption of information processing at the cortical level due to the loss of the focusing action of subcortical inputs.

8.4. Orbito-frontal cortex (OFC)

The OFC is a large brain region, which encompasses both rostral and ventromedial areas. Because, it receives multimodal inputs from the temporal association cortex, amygdala and hypothalamus as well as limbic components of the basal ganglia, it has been viewed as the highest integration center for emotional processing (Krawezyk, 2002). By analogy with the Dorsolateral prefrontal cortex (DLPC), which is the prefrontal area for parietal lobe, the OFC can be regarded as the prefrontal area for the temporal lobe. The OFC seems to play a role in situations involving incentives/bonus/rewards and in conditions, where the subject has to make rapid alterations in behavior to accommodate the environmental changes. Several lines of evidence suggested that OFC played a crucial role in the decision-making process based on rewards. Patients with orbitofrontal damage experience great difficulties in decision-making. They also tend to take risks, whether profitable or not (Miller, 1985). Experimental lesions of OFC in monkeys have shown impairment in reward–related learning tasks and irrespective of the nature of the sensory context. These monkeys also exhibited an absence of emotional reaction to environmental stimuli. OFC neurons become particularly active, when the animal is placed in a situation where, it expects and receives a reward. Interestingly, face selective neurons have been reported in OFC, which may be relevant in the detection of facial expression which is a critical point in social decision. These neurons may be involved in the association between a positive reward value and a particular facial expression. The OFC seems to play a predominant role in motivational aspects of decision-making. Among the more posterior cortical areas, the left inferior parietal cortex and parieto-occipital junction are involved in cognitive tasks related to visual imagery. The underactivity of these regions could probably explain the spatial memory deficits and visual memory deficits observed in OCD patients. The repetitive rituals (compulsions) and aggressive behavior, which is predominant in OCD patients is probably due to serotonin depletion.

8.5. Striatum

The striatum is known to be formed by two types of information-processing modules: the striasomes and the matrisomes. The striasomes receive information from the limbic structures such as amygdala, OFC & ACC (Eblen and Graybiel, 1995). In turn, they send projections to the dopaminergic neurons of the substantia nigra. These anatomical findings suggest that the striasomes could also play a role in the emotional modulation of cortical information. The matrisomes receive information from the lateral parts of the premotor and prefrontal cortices, which are involved in the anticipation behavior and planning (Flaherty and Graybiel, 1994). The cholinergic inter-neurons of the striatum, viz. tonically active neurons (TANS), may be playing a particular role in integrating the information flowing through the striasomes and matrisomes. These neurons could constitute a neural system that is involved in the processing of several aspects of information, such as the detection of unpredicted events or the context of stimulus discrimination (Ravel et al., 2001).

The limbic part of the striatum (ventral striatum) under the control of the dopaminergic afferents might be involved in reward–driven learning processes. On the other hand, the dorsal striatum seems to be involved in the procedural learning of behavioral routines that are performed almost without conscious effort. In particular, in the context of procedural learning, the disruption of the "readiness" and "release" functions ascribed to the striatum might support some aspects of OCD pathophysiology. However, the striatum could also play a part in other processes potentially disrupted in OCD, such as emotional modulation of information and representation of the expected consequences of action. On the other hand, the performance of repetitive behaviors in OCD patients might have a positive effect on the reduction of anxiety, a process that can be assimilated to some form of reward.

8.6. Amygdala

In the past decade, much research has been focused on the neural substrates that are involved in the expressions of fear and anxiety. The amygdala and its various outputs might play a major role in mediating the clinical signs and symptoms of fear and anxiety (Le Doux, 2000). Schematically, the amygdala is comprised of several nuclei, such as the lateral nucleus, basolateral nucleus and central nucleus. However, recent evidence supports the fundamental idea that the amygdala is not only involved in negative emotions such as fear and anxiety but also in reward and motivational processes through reciprocal connections to the nucleus accumbens and the OFC. Thus, the amygdala appears to play an important role in the expressions of emotion and motivation, probably through its connections with the OFC, ACC and ventral striatum. A dysfunction of this structure, as suggested by some neuroimaging studies in OCD patients, might mediate the non-specific anxiety experienced relative to obsessive thoughts.

8.7. Thalamus

The diencephalic position of the thalamus in the brain explains why it receives large cortical inputs. It participates in emotional expression through the AN (anterior nucleus of the thalamus), which is connected to the MB (mammillary bodies) and, in turn, sends projections to the ACC. The putative role of the VA (ventral anterior nucleus of the thalamus) in cognitive functions involving attention and working memory is based on the link with DLPC. Discrete parts of the MD (medial dorsal nucleus of the thalamus) seem to be important in both emotional and cognitive processing through their preferential anatomical connections with the OFC and DLPC. Thalamic dysfunction has been associated with deficits in executive functions like planning, goal directed behavior, attention, and working memory (Lacerda et al., 2003).

8.8. Brainstem inputs

The mesocorticolimbic dopaminergic system emanates from the ventral mesencephalon, which encompasses the Ventral tegmental area (VTA), and projects to the nucleus accumbens with other limbic ventral striatal regions and cortical areas, especially the OFC, DLPC, ACC. The effects of the lesions, receptor blocking agents, electrical stimulation and self-administration of drugs of abuse suggest the effective contribution of the mesocorticolimbic system in the attentional, emotional and motivational processes. Dopamine contributes to the organization and regulation of goal-directed behavior. The serotonin-producing neurons are mainly located in the brainstem raphe nuclei. The description of the anatomy and development of the brainstem raphe nuclei has shown that they form the largest and most complex neurochemical efferent system in the human brain. General theories have attributed a broad range of behavioral functions to serotonin, which is considered as a general inhibitor of motor behavior. In contrast, reduced serotonin function has been shown to increase exploration, locomotor activity, aggression and sexual behaviors in animals and human beings (Lucki, 1998). The repetitive rituals (compulsions) and aggressive behavior, which is predominant in OCD patients is probably due to serotonin depletion.

- PET Scans show people with OCD have different brain activity from others
- Another theory: miscommunication between the orbital frontal cortex, the caudate nucleus and the thalamus
- Caudate nucleus doesn't function properly and causes thalamus to become hyperactive and sends never-ending worry signals between OFC and thalamus OFC responds by increasing anxiety

The Group A Beta–Hemolytic Streptococcal (GABHS) infections produce anti-neuronal antibodies that adversely affect basal ganglia cells, which might lead to the development of OCD (Swedo et al., 1998). The susceptibility marker that may predispose some individuals to

http://dx.doi.org/10.5772/57193

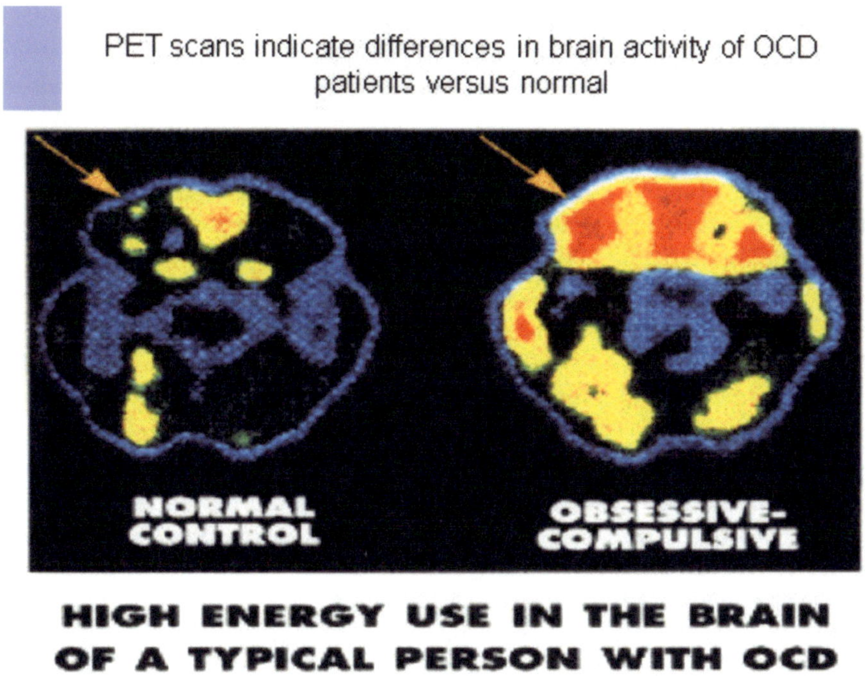

Figure 2. High energy use in the brain of a typical person with OCD

develop OCD has been well identified. D8/17 ((the antigen present on the surface of peripheral blood mononuclear cells) positive individuals develop OCD as a consequence of their autoimmune response to Group A beta-hemolytic streptococcal infection, a response that is believed to yield antibodies which cross react with basal–ganglia antigens and produce tissue damage. The orbitofrontal cortex that had been demonstrated to be overactive in OCD is a region mediating the active expression of emotional response to significant biological stimuli as well as the inhibition of behavioral response.

9. Treatment of OCD

9.1. Pharmacological treatments

The serotonin reuptake inhibitors (SRIs) are consistently effective in patients of Obsessive-compulsive disorder. The anti-obsessional effects of potent SRIs produce progressive desensitization of the presynaptic autoreceptors present on 5HT neurons and their nerve terminals, thereby increasing synaptic 5HT release in the orbitofrontal cortex. Clomipramine was the first to show beneficial effects on OCD symptoms. The newer generation of antidepressant drugs viz. fluvoxamine, fluoxetine, paroxetine, sertraline and citalopram have also been found useful in management of OCD. The mean daily dosage is 50–200 mg for sertraline, 20–80 mg for fluoxetine, 40–60 mg for paroxetine and 150–300 mg for fluvoxamine. The atypical antipsychotics such as risperidone act as new therapeutic options for refractory OCD. Marble burying behavior of mice has been used to model anxiety disorder including obsessive–compulsive disorder (OCD) due to the excessive nature of the behavior and due to the pharmacological effects of clinical standards. Gaikwad et al., 2007 suggested that LHRH agonist such as leuprolide may be clinically effective in OCD, resulted dose dependently attenuated marble-burying behavior in mice has been used to model anxiety disorders viz. obsessive–compulsive disorder. Acute administration of the neurosteroid allopregnanolone (i.c.v.) or its precursor, progesterone, decreased marble burying in male mice, but administration of the 5-reductase inhibitor finasteride, an allopregnanolone indirect antagonist, did not affect marble burying (Umathe et al., 2009). In addition, acute administration of the 5-HT2a/2c antagonist ritanserin abolished the leuprolide-induced decrease in marble-burying of male mice without having any effect on locomotion. Ritanserin treatment on its own had no effect on marble-burying or on locomotion (Gaikwad et al., 2010). These results suggest that the anti-compulsive effects of GnRH may depend upon serotonergic activity, and specifically, that the effects of leuprolide in the marble-burying model are mediated through 5-HT2A/2C receptors (Gaikwad et al., 2010).

9.2. Neurosurgical treatments

Neurosurgical treatments have been used for the management of chronic, severely distressing forms of OCD where conventional treatments are ineffective. In the United States and Canada, anterior cingulotomy was the most widely used neurosurgical procedure applied in the treatment of anxiety refractory OCD. Two surgical techniques have been used: radiofrequency thermolesion or thermocapsulotomy and the newer radiosurgical or gamma knife capsulotomy techniques. It has been shown that 50–70% of patients with OCD respond favorably to this type of operation at the end of the follow-up period ranging from 1 to 9 years. The Bilateral thermolesions are produced by radiofrequency in the anterior cingulate cortex (ACC) (Broadman areas 24 and 32). It has found that 25% of patients with OCD were slightly improved, 31.3% were markedly improved, 12.5% were functionally normal on medication or psychotherapy maintenance and 12.5% were essentially well without any treatment at the mean follow-up of 9 years. The Y-BOCS used for assessment of OCD symptom severity, resulted a moderate-to-marked improvement (defined as a 50% or greater reduction in Y-BOCS scores) in 57% of cases at the mean 13-year follow-up. Approximately 50–70% of patients with OCD showed good outcome by surgery, indicated no or mild residual symptoms at the follow-up ranging from 16 months to 8 years. The recent report suggested that good outcome was observed in only 33% of cases within 1 year of surgery. In the United Kingdom limbic leucotomy was developed, a technique based on making bilateral lesions of the cingulotomy, in addition to those of the original subcaudate tractotomy.

9.3. Psychological treatments

Psychological treatments based on cognition-behavioral therapy (CBT) are effective in the treatment of OCD, alone or in combination with SRIs. The current procedures are in vivo (real-life) exposure with self-imposed response prevention (EX/RP) which showed beneficial effects of this behavioral approach in two patients with severely disabling OCD. The reduction in OCD symptom severity was found after 3–7 weeks of EX/RP treatment while no change occurred during the control condition. The goal of CBT treatment is to learn to respond appropriately to intrusive thoughts and urges of OCD in new and much more adaptive ways and to reattribute the belief about "false brain messages" as intimately due to a biomedical disease state. The therapeutic effects showed the functional changes in interactions between the limbic cortex (including the orbital cortex and anterior cingulate gyrus) and the basal ganglia that are extremely important in redirecting information flow toward the integration of new behaviorally significant environmental events during learning. This function may be central in the acquisition of more adapted patterns of behavioral responses during the course of CBT.

10. Conclusion

OCD is an anxiety disorder featuring intrusive and troubling thoughts, which are perceived as the products of one's own mind unlike schizophrenia. The Patient affected by OCD feels compelled to carry out certain stereotyped behaviors, although he recognizes that his behavior is at times irrational. Entire brain functioning is disturbed in patients suffering from OCD, thereby producing devastating effects at the work-place as well as at homes of the patients. OCD is a complicated disorder. Selective serotonin reuptake inhibitors (SSRIs) and to some extent tricyclic antidepressants form the main stay in the symptomatic treatment of OCD. Most of the OCD cases are incurable. Therefore, there is a great challenge to discover new drug treatment for the management of OCD.

Author details

Uday Gaikwad

Department of Pharmacology, R.D's College of Pharmacy, Pune (Maharashtra), India

References

[1] Baer, L. (1993). Behavior therapy for obsessive compulsive disorder in the office based practice, J. Clin. Psychiat., 54 suppl , 6, 10-5.

[2] Bergqvist, P. B, Bouchard, C, & Blier, P. (1999). Effect of long-term administration of antidepressant treatments on serotonin release in brain regions involved in obsessive-compulsive disorder, Biol. Psychiatry., 41, 64-174.

[3] Cruz, C, Camarena, B, King, N, Paez, F, Sidenberg, D, De La Fuente, J. R, & Nicolini, H. (1997). Increased prevalence of the seven-repeat variant of the dopamine D4 receptor gene in patients with obsessive-compulsive disorder with tics, Neurosci. Lett., 23, 1-4.

[4] David, W. E, Marc, D. L, & Emily, I. (2004). The role of the orbitofrontal cortex in normally developing compulsive-like behaviors and obsessive compulsive disorder, J. Brain & cogn., 55, 220-234.

[5] Enoch, M. A, Rotondo, A, Kaye, W. H, Murphy, D. L, Goldman, D, & Greenberg, B. D. (1998). HT_{2A} promoter polymorphism 1438 G/A, anorexia nervosa, and obsessive-compulsive disorder, Lancet., 351, 1785-1786.

[6] Goodman, W. K, Ward, H, Kablinger, A, & Murphy, T. (1997). Fluvoxamine in the treatment of obsessive-compulsive disorder and related conditions, J. Clin. Psychiatry., 58 Suppl , 5, 32-49.

[7] Greist, J. H, Jefferson, J. W, & Kobak, K. A. (1995). Efficacy and tolerability of serotonin transport inhibitors in obsessive-compulsive disorder: A meta- analysis, Arch. Gen. Psychiat., 52(1), 53-60.

[8] Gaikwad, U, Bhutada, P, Wanjari, M, & Umathe, S. (2007). LHRH antagonist attenuates the effect of fluoxetine on marble-burying behavior in mice. *Eur. J. Pharmacol., 563*, 155-159.

[9] Han, L, Nielsen, D. A, Rosenthal, N. E, Jefferson, K, Kaye, W, Murphy, D, Altemus, M, Humphries, J, Cassano, G, Rotondo, A, Virkkunen, M, Linnoila, M, & Goldman, D. (1999). No coding variant of the tryptophan hydroxylase gene detected in seasonal affective disorder, obsessive compulsive disorder, anorexia nervosa and alcoholism, Biol. Psychiat., 351, 1785-1786.

[10] Leckman, J. F, Rasmussen, S. A, Peterson, B. S, Grice, D. E, Boardman, J, Zhang, H, Vitale, A, Cohen, D. J, Bondi, C, Alsobrook, J, Goodman, W. K, Mc Dougle, C. J, & Pauls, D. L. (1997). Symptoms of obsessive compulsive disorder, Am. J. Psychiat., 154, 911-917.

[11] Mataix-cols, D, Rasmussen, S. A, Goodman, W. K, Rauch, S. L, Baer, L, Jenike, M. A, Eisen, J. L, & Shera, D. M. (2002). Symptom stability in adult obsessive compulsive disorder: data from a naturalistic two-year follow up study, Am. J. Psychiat., 59, 263-268.

[12] Mcdougle, C. J, Epperson, C. N, Pelton, G. H, Wasylink, S, & Price, L. H. placebo-controlled study of risperidone addition in serotonin reuptake inhibitor-refractory obsessive-compulsive disorder, Arch. Gen. Psychiatry., 57, 794-801.

[13] Mcdougle, C. J, Goodman, W. K, & Price, L. H. (1994). Dopamine antagonists in tic-related and psychotic spectrum obsessive compulsive disorder, J. Clin. Psychiatry., 55 Suppl, 24-31.

[14] Millet, B, Chabane, N, Delorme, R, Leboyer, M, Leroy, S, Poirier, M. F, Bourdel, M. C, Mouren-simeoni, M. C, Rouillon, F, Loo, H, & Krebs, M. O. (2003). Association between the dopamine receptor D4 (DRD4) gene and obsessive-compulsive disorder, Am. J. Med. Genet., 116B, 55-59.

[15] Mundo, F, Richter, M. A, Sam, F, Macciardi, F, & Kennedy, J. L. (2000). Is the 5- HT (1D beta) receptor gene implicated in the pathogenesis of obsessive compulsive disorder? Am. J. Psychiat., , 157, 1160-1161.

[16] Pato, M. T, Pato, C. N, & Pauls, D. L. (2002). Recent findings in the genetics of OCD, J. Clin. Psychiatry., 63 Suppl , 6, 30-33.

[17] Rapoport, J. L, Leonard, H. L, & Swedo, S. E. (1993). Obsessive compulsive disorder in children and adolescents: issues in management, J. Clin. Psychiat., 52 suppl, 6, 27-9.

[18] Schindler, K. M, Richter, M. A, Kennedy, J. L, Pato, M. T, & Pato, C. N. (2000). Association between homozygosity at the COMT gene locus and obsessive-compulsive disorder, Am. J. Med. Genet., 96, 721-724.

[19] Stein, M. B, Forde, D. R, Anderson, G, & Walker, J. R. (1997). Obsessive compulsive disorder in the community: an epidemiologic survey with clinical reappraisal, Am. J. Psychiat., 154, 1120-1126.

[20] Zohar, J, Mueller, E. A, Insel, T. R, Zohar-kadouch, R. C, & Murphy, D. L. (1987). Serotonergic responsivity in obsessive-compulsive disorder. Comparison of patients and healthy controls, Arch. Gen. Psychiatry., 44, 946-951.

[21] Bush, G, Luu, P, & Posner, M. I. (2000). Cognitive and emotional influences in anterior cingulated cortex, Trends. Cogn. Sci., 4, 215-222.

[22] Eblen, F, & Graybiel, A. M. (1995). Highly restricted origin of prefrontal cortical inputs to striosomes in the macaque monkey, J. Neurosci., 15, 5999-6013.

[23] Flaherty, A. W, & Graybiel, A. M. (1994). Input-output organization of the sensorimotor striatum in the squirrel monkey, J. Neurosci., 14, 599-610.

[24] Krawezyk, D. C. (2002). Contributions of the prefrontal cortex to the neural basis of human decision making, Neurosci. Biobehav. Rev., 26, 631-664.

[25] Lacerda, A. L, Dalgalarrondo, P, Caetano, D, Hass, G. L, Camargo, E. E, & Keshvan, M. S. (2003). Neuropsychological performance and regional cerebral blood flow in obsessive compulsive disorder, Prog. Neuropsychopharmacol. Biol. Psychiat., 27, 657-665.

[26] Le DouxJ.E. ((2000). Emotion circuits in the brain, Annu. Rev. Neurosci., 23, 155-184.

[27] Lucki, I. (1998). The Spectrum of behaviors influenced by serotonin, Biol. Psychiat., 44, 151-162.

[28] Miller, F. K. (1999). The prefrontal cortex: complex neural properties for complex behavior, Neuron., 22, 15-17.

[29] Miller, L. (1985). Cognitive risk-taking after frontal or temporal lobectomy. The synthesis of fragmented visual information, Neuropsychologia., 23, 359-369.

[30] Ravel, S, Sardo, P, Legallet, E, & Apicella, P. (2001). Reward unpredictability inside and outside of a task context as a determinant of the responses of tonically active neurons in the monkey striatum, J. Neurosci., 21, 5730-5739.

[31] Saxena, S, Brody, A. L, Schwartz, J. M, & Baxter, L. R. (1998). Neuroimaging and frontal-subcortical circuitry in obsessive compulsive disorder, Br. J. Psychiat., Suppl, 26-37.

[32] Swedo, S. E, Leonard, H. L, & Garvey, M. (1998). Pediatric autoimmune neuropsychiatric disorder associated with streptococcal infections: clinical description of the first so cases, Am. J. Psychiat., 155, 264-271.

Chapter 2

Cognitive Processes and Biases in Obsessive-Compulsive Disorder

Elisa Krackow, Sarah Nunley and Pamela Tessier

Additional information is available at the end of the chapter

http://dx.doi.org/10.5772/58406

1. Introduction

Empirically-supported treatments for obsessive-compulsive disorder (OCD) emphasize the behavioral principles of exposure and response prevention [1]. However, theory and data suggest that cognitive components play a substantial role in the etiology and maintenance of OCD. This chapter will review cognitive theories and models of OCD, and cognitive processes in OCD including thought-action-fusion, inferential confusion, reality monitoring, cognitive inhibition, and memory and attentional biases. A comprehensive treatment that draws on both the cognitive literature reviewed herein plus exposure and response prevention is proposed.

2. General cognitive theories of OCD

The seminal cognitive theory by Rachman describes a diathesis-stress model in which an individual's history of being reared in an environment where strong values are placed upon the person's thoughts increases vulnerability to OCD [2]. Current levels of stress and negative affect increase the likelihood of obsessions which tend to be elicited by cues in the external environment [3]. Learning history then increases the likelihood of the obsessions being appraised as significant or meaningful. The appraisal of the obsessions as significant represents a cognitive bias because without a cognitive vulnerability towards OCD, credibility would not be given to the thoughts that become obsessions. These processes are more likely to occur in individuals with pre-existing trait anxiety. For example, an anxious person may read a story in a newspaper about an electrical house fire and then obsess over the possibility of accidentally setting an electrical fire.

It is the above combination of factors that lays the groundwork for obsessions, coupled with other specific cognitive processes and biases which are about to be discussed. Overestimating the probability of the occurrence of negative events can occur if one believes that one is responsible in a situation. For example, a person may believe that "a child is more likely to have an accident if I am babysitting than if someone else is babysitting." Relatedly, Rachman discusses the cognitive process of Thought-Action Fusion (TAF) [2]. TAF is a cognitive distortion representing a moral component and an event probability component. A moral reasoning error is made when an individual has a thought about an action and then appraises that thought as being equivalent or almost equivalent to actually having engaged in the action itself. In TAF, the probability of an event occurring can be perceived as increasing simply by having a thought about the event. Both the moral reasoning and event probability components fuse thoughts and actions. TAF is increased by the overestimation of attributions of responsibility to the self for events, which represents a cognitive distortion. It should be noted that although TAF is a component of Rachman's cognitive theory, it is described in other literature as a separate theory [4].

Salkovskis (1985) proposed a second cognitive theory of OCD [5]. As Salkovskis points out, the components of this theory have substantial overlap with Rachman's previously described theory. According to this theory, obsessions can be cued via stimuli in the environment. Cuing of obsessions is aversive to the individual and therefore avoidance measures are taken. The person may or may not recognize the obsessional thought to be irrational. If the thought does not have negative ramifications for the person, this cognitive chain of events will end. If negative ramifications are present, automatic thoughts, similar to Beck's conceptualization of automatic thoughts will occur [6]. Automatic thoughts are more likely to occur if the current mood state is negative. These thoughts interact with a set of assumptions, including but not limited to, thought equals action, increased sense of responsibility for events, beliefs about the need to control thoughts, and self blame for events. These negative thoughts then influence mood which leads to attempts to reduce distress by engaging in compulsions. Negative reinforcement occurs when the distress is reduced lending credibility to the negative automatic thoughts. Co-occurring depressed mood can also play a role via several mechanisms. The first mechanism increases the dysfunctional beliefs via increasing the stimuli that may cue an intrusive thought that could become an obsession. This can then create a transactional process by which mood increases negative automatic thoughts and in turn, these thoughts then can increase negative mood.

Rachman proposed a specific cognitive theory of the compulsive checking subtype of OCD [7]. According to this theory, individuals with this subtype of OCD have an increased belief in self-responsibility for maintaining the safety of self and others. The increased sense of responsibility comes from an unrealistic appraisal of the probability of harm. Repeated checking serves the function of increasing certainty that any possible threat will be alleviated. Given that certainty is improbable, checking continues. Repeated checking results in memories of repeated events becoming fuzzy, and therefore confidence in memory as to whether checking has or has not occurred is decreased. That is, checking behavior temporarily reduces the anxiety via engaging in the compulsive behavior itself or via negatively reinforcing cognitions.

3. Specific cognitive processes in OCD

Thought-action fusion and OCD. Thought-action fusion is a component of OCD in the two previously discussed theories [2,5]. Researchers have developed assessment measures of TAF which have resulted in empirical findings supporting TAF as a process in OCD. Most commonly, TAF is measured using the Thought-Action-Fusion questionnaire, a 19-item questionnaire composed of moral TAF, likelihood-self TAF, and likelihood-other TAF scales [8]. Accordingly, the two factors of moral and likelihood TAF accounted for 71% of the variance. Internal consistency ranged from.88 to.96, and the TAF converged as expected with the Maudsley Obessive Compulsive Inventory. As expected, an OCD sample exhibited higher TAF scores on all three scales than did two pathology-absent adult comparison samples. TAF is not specific to patients with OCD but rather occurs with other anxiety disorders [9].

Researchers have examined whether TAF is more common to depression than to OCD due to high negative affect being characteristic of depression [10]. Drawing on Rachman's model, Abramowitz and colleagues theorized that negative affect mediates the relationship between TAF and OCD [2]. Their results supported this view. Interestingly, this study also found that the likelihood component but not the moral component of TAF is present in individuals with OCD. Researchers have further specified the relationship between these variables [11]. They examined the relation between TAF, schizotypal traits (magical thinking), OCD symptoms, general anxiety, and depression. Schizotypal traits were deemed important to include because nearly half of people with OCD have schizotypal personality characteristics. Among those characteristics are magical thinking, which the authors note may be associated with cognitive biases and therefore increased vulnerability to OCD. Schizotypical traits were associated with likelihood TAF such that the relation between TAF and OCD disappeared when schizotypal traits were controlled. No significant relation was found between OCD and moral TAF. In a third investigation, other investigators examined TAF, self-doubt regarding the belief that the compulsion will fix the problem, and cognitive control when obsession related beliefs emerge in youth with OCD, youth with another anxiety disorder, and a psychopathology-absent comparison group [12]. They found that higher likelihood TAF occurs in youth with OCD compared to youth without psychopathology, with no significant differences between the OCD and other-anxiety disorder groups, but self-doubt does not differ between the OCD and control group. Moral TAF was not measured in this study.

In sum, people with OCD tend to make cognitive errors in their thinking with regard to the fusion of thoughts and actions. This tends to occur in the following two ways: 1). Having a thought can be viewed as being as "bad" as having completed the action (moral TAF) and 2). thinking about the events increases the perception that the event will occur (likelihood TAF). The evidence appears to be slightly stronger for likelihood TAF than for moral TAF, but moral TAF was not measured in all studies examining this construct. Some evidence suggests that TAF is not specific to OCD but instead also occurs in anxiety disorders in general.

Inferential Confusion and OCD. A concept related to TAF is "inferential confusion" or the process of confusing imagined events and actually occurring events. According to this view people with OCD mistake the imagined negative event for a real event, such that the imagined event becomes more realistic creating the belief that the event has an increased probability of occurrence [13]. Devaluation of information taken in by the senses (e.g., the tendency to believe something despite the available evidence to the naked eye) and inferential reasoning errors combine to make conditions ripe for the confusion of real and imagined events [14]. A "crossover point" can be identified in these individuals in which obsessions begin and can be identified [13]. Compulsions result from the obsessions as attempts to modify the probability of the event occurrence. Engaging in compulsions gives credibility to the imagined event. A specific type of inferential confusion can occur in which thinking a particular thought makes it likely that a person will act on that thought. Therefore, similar to TAF, a thought is equivalent to a personal action.

Support for inferential confusion theory comes from a follow-up study in which participants with OCD were given questionnaires measuring inferential confusion and obsessional thinking [15]. Analyses showed strong overlap between inferential confusion and common obsessions such as contamination and harm, and compulsions, such as washing. Even when other variables were controlled (e.g., overestimation of threat and general anxiety), the relationship between inferential confusion and obsessions/compulsions remained. Additional data show that when inferential confusion improves (as measured by a specific paper and pencil questionnaire of inferential confusion), OCD symptoms improve [16].

In a nonclinical sample, schizotypal symptoms (delusional thinking, perceptual disturbances, living in an inner world) and inferential confusion predicted obsessive compulsive behavior, with the majority of the predictive validity coming from the perceptual disturbances subscale of the schizotypal measure (22% of the variance) [17]. Inferential confusion explained 5% of the variance. The authors note that people with OCD typically do not have perceptual disturbances, but perceptual disturbances can be considered to be a product of inferential confusion.

Reality monitoring and intrusive imagery in OCD. Reality monitoring errors occur when people confuse real and imagined events [18]. It has been suggested that poorer reality monitoring ability in OCD individuals due to intrusive imagery present in OCD [19]. That is, with the occurrence of repeated involuntary images, those images seem real which leads participants to confuse them with reality. Indeed there is a large body of literature now showing that imagination leads people to believe in events that never occurred (see [20] for a review). Therefore, other researchers examined reality monitoring ability for actions in an inpatient group with OCD and a yoked nonanxious control group [21]. In this study, participants were given sequences of 6 actions and instructed to either perform an action or imagine performing an action. Action sequences were either anxiety provoking or neutral. There were no significant differences in ability to identify imagined versus performed actions between the groups. Results did not differ as a function of whether the action sequences were anxiety-provoking or neutral. However, people with OCD had less confidence in their reality monitoring ability.

Relatedly, deficits in reality monitoring ability have been shown to occur in people with subclinical OCD symptoms (specifically checking behaviors) compared to people without OCD symptoms [22]. To document this effect, all participants were read a list of actions. They were then asked to perform an action, observe an action being performed on a videotape, or write down a never performed action that was read to them by an experimenter. Following this task, participants were then asked to generate a written list of all actions they could recall, and then reality monitor by indicating whether they had performed, observed, or written each action they had recalled. People with OCD symptoms recalled fewer of the performed, written, and observed actions. In addition, they miscategorized (made more reality monitoring errors) whether they performed an action, wrote down the action, or observed the action being performed.

Additional data suggest that intrusive imagery related to the experience of adverse life events is a component of OCD at least in some patients. In an inpatient OCD sample, 81% of 34 patients reported intrusive images directly from or related to adverse life events [23]. Interestingly, none of the patients met co-morbid criteria for PTSD. OCD patients with intrusive images had higher levels of general anxiety and more OCD symptoms than OCD patients who did not have images. However, the number of patients without images was small (*n*=7). Consistent with Rachman's previously discussed theory, patients who reported imagery felt greater responsibility for events than those who did not report images, although the origins of this increased responsibility are unclear.

Other reports of intrusive images in OCD have been found in the literature [24]. Therefore, Rachman correlated a single item measuring intrusive images with mental contamination (feeling dirty inside) [24]. This item correlated approximately.54 with other measures of OCD. Based on analyses from another archival dataset, Rachman reported that mental imagery items correlated.34-.445 with other measures of OCD [24]. Therefore, it has been suggested that imagery rescripting of distressing events might be an intervention appropriate for these individuals, but also point out that it is unknown whether this would decrease OCD symptoms given the lack of current empirical data for this intervention for OCD [23].

In sum, the evidence suggests that imagery can be an important in the treatment of people with OCD who experience intrusive images from negative life events. More broadly, imagery plays a role in the development OCD. Imagery acts as a mechanism by which obsessions come to seem more realistic thereby increasing the perception of the likelihood of occurrence. However, specific tests of reality monitoring ability do not always show differences in the performance of people with and without OCD. Methodological issues, including the timing of memory tests (i.e., lack of a substantial delay between exposure to the stimuli and memory test), may have influenced the results of existing studies. Therefore, further research is needed.

4. Memory biases in OCD

Negative memory biases have been theorized to play a role in psychological disorders, particularly anxiety. According to Beck, negative schemata formed in childhood render an individual susceptible to drawing attention towards threatening stimuli which is theorized to positively impact memory for these stimuli [6]. These schemata are activated when ambiguous or anxiety-provoking stimuli are encountered and predispose a person towards an anxious interpretation of events. A more specific model of how anxiety influences the allocation of attention and memory has been developed [25]. Specifically, when a stimulus is encountered in the environment, it is mapped onto existing schemata and a decision is made as to the threat level of the stimulus. When stimuli are appraised as threatening, they receive further cognitive resources allocated towards them. That is, attention is drawn towards those stimuli, particularly in people high in trait anxiety. The opposite cognitive allocation or drawing attention away from the stimulus occurs in people low in trait anxiety which serves the function of not allocating unnecessary cognitive resources in the case of stimuli that do not warrant attention. Williams and colleagues discuss how the allocation of attention is a first step towards impacting or improving memory for those threatening stimuli [25].

Despite these theories, the rationale for studies examining memory biases in OCD focuses on studying phenomenon that have been observed in clinical settings rather than theory-based predictions. For example, because it has been noted that people with OCD express doubt regarding whether they have performed a particular behavior (e.g., checked whether they turned the stove off) which might lead to repeating the compulsion. One study compared people with the contamination subtype of OCD to both anxious controls to an undergraduate sample (n=10 per group in the OCD and anxious control conditions) [26]. In this study, participants watched an experimenter touch objects with a "dirty" tissue (i.e., contaminate) or touch objects with a "clean" tissue (i.e., not contaminate). Memory was tested immediately following the task. People with OCD remembered "dirty" objects better. This pattern of findings did not hold for both the anxious control group or the undergraduate comparison sample. It should be noted that items were scored only for correct recall in this study. Analyses did not include errors in recall.

Other researchers replicated the above study [26] using a sample that included different control groups and included an analysis of recall errors [27]. The study included 16 OCD washers, 16 OCD checkers, 16 controls with anxiety (specific phobia), 16 in a no-pathology comparison group. Participants entered a room with 50 items which they then observed the experimenter touch with a tissue (prevents contamination) or bare hand (brings into contact with contamination). Participants were asked to recall the items 50 minutes later by first generating a written list of all items they recalled. For the recalled items only, they were then asked to recall whether the experimenter touched each item with tissue or bare hands. No significant group differences emerged in total number of items recalled in free recall. There was a difference in recall of whether the object was touched by the experimenter with tissue vs. hand. OCD washers showed superior accurate recall as to whether the objects were touched by bare hands. The anxious control group showed superior recall for objects touched by a tissue. These results suggest superior memory in OCD washers for "contaminated" items.

Another study examined whether differences in recall existed between an OCD sample and control sample as a function of the anxiety provoking properties of the to-be-remembered experienced versus imagined actions [21]. Participants had a booklet with 6 actions listed. They were asked to perform those actions. They were then taken to a room and were asked to either imagine or perform a set of 20 actions. They were then asked to state whether only the last item was imagined or performed. Following this they were asked for a series of ratings of vividness, confidence, desire for vividness, and satisfaction with memories. Results showed no significant difference in reality monitoring performance across anxiety provoking and nonanxiety provoking stimuli, but a significant difference in desire for vividness emerged and was greater in the OCD group. No significant differences were found on the other 3 ratings. However, the sample size in this study was small and may not have been sufficient to detect significance (*N*=19; 12 OCD, 7 controls).

McNally and colleagues examined directed forgetting in OCD patients and controls [28]. Participants were presented with words of a negative, positive, or neutral valence followed by instructions directing them to remember or forget the just-presented stimuli. OCD patients had a more difficult time forgetting negative words than did comparison participants. However, it should be noted that a small portion of the words seemed to be directly related to specific types of OCD (contamination) and it is unclear whether the results would hold if these items were removed. It would be important to compare an OCD relevant negative list to another non-OCD related negative list to determine whether the above findings hold.

Another study examined memory as a function of experimenter induced sense of responsibility [29]. The authors note that people with OCD report a strong sense of responsibility for outcomes including correctness of their checking or lack of checking behavior. For example, one might think, "If I didn't look at the stove carefully to see whether it was off, it might still be on and the house may catch on fire." Following a baseline check in their homes to induce anxiety, the researchers had participants with OCD (all handwashers *N*=11) complete their typical checking behaviors (e.g., check to make sure the stove is off) in their homes under conditions of low and high responsibility. In the high responsibility condition, participants were told that they were responsible for completing the check without assistance and were responsible for the accuracy of the check. They were asked to sign a contract to this effect agreeing to these terms. In the low responsibility condition, participants signed a contract that attributed the responsibility of the check to the research assistant. Participants completed a memory interview immediately after the checks and provided confidence ratings for their responses to the interview questions. The memory interview included memory for specific aspects of the check that might have been perceived as threatening (" What was the final state of the thermostat?" (Radomsky et al., 2001, p. 817) [29] as well as nonthreatening aspects of the check ("Did I (the experimenter) cough during the check?) (Radomsky et al., 2001, p. 817) [29]. In the next phase of the study, all participants returned to the lab a week later and watched videos of both checks and then completed the same memory interview and confidence ratings. This delayed phase represented a no responsibility phase. Results showed no overall differences in the amount remembered across the conditions, but under conditions of high responsibility, OCD handwashers recalled more threatening information than irrelevant information. This result did not hold under conditions of low responsibility. When the people who were originally in the high responsibility group were tested under conditions of no responsibility one week later, they no longer showed superior recall of threatening information. However, this result is confounded by methodological issues. That is, threatening actions were sometimes performed by the participant whereas nonthreatening actions were sometimes performed by the experimenter. There is now a large body of literature showing superior memory for participation versus observation [30].

One study measured explicit memory with a recognition task for previously heard or never heard sentences and confidence [31]. Implicit memory was assessed via participants' ratings of the volume of the previously heard or never heard sentences. There was no tendency for OCD participants to rate previously heard sentences as louder thereby demonstrating a lack of a superior implicit memory bias. As expected, OCD participants did not show an explicit memory bias; that is they did not exhibit superior recognition performance compared to control participants nor did they correctly reject never-presented sentences at higher rates than control participants. Instead, OCD patients rated all sentences as louder than matched nontreatment seeking controls suggesting what the authors referred to as a "perceptual deficit," or difficulty accurately perceiving the stimuli, rather than memory deficit, although the ramifications of this finding are unclear. Failure to find an explicit memory bias appeared to be due to high accuracy rates in both groups. Perhaps a longer test delay would have been beneficial in this study.

In sum, the data on memory biases in OCD show that the majority of studies fail to find evidence for such as bias. Numerous methodological issues in the above described studies exist, thereby precluding firm conclusions from being drawn. Small sample size is clearly one limitation in some studies examining this topic. It is also possible that delay of test was not long enough in some of the above studies for differences to emerge. In some studies, participants were informed prior to encoding that memory would be tested, thereby raising the possibility that rehearsal strategies were used following experimenter notification that a memory test would occur. Other methodological issues include failure to score incorrect recall and confounds regarding the effects of participation.

5. Attentional bias in OCD — Theory and research

According to theory, individuals are more vulnerable to emotional disorders when the level of emotion exceeds an individual's capacity to control that emotion [32]. Individuals with anxiety disorders are more likely to display selective attention to threatening stimuli and experience more difficulty disengaging once attention to threatening stimuli is activated. It is also the case that the threshold for appraising a stimulus as threatening is lower in people with anxiety disorders. Accordingly, attention towards threat is a combination of current level of state anxiety, level of threat appraisal, and ability to modulate attention. Mathews and MacLeod suggest that attentional bias plays a role in the development of anxiety disorders [32]. Although many studies show that people with anxiety disorders display an attentional bias, these data do not show that attentional bias causes anxiety disorders. Indirect support for the hypothesis that attention plays a causal role in anxiety comes from studies that demonstrate that creating cognitive bias impacts emotional state. It is important to acknowledge that to date there has not been a study conducted in which an attentional bias towards anxiety is created with the resulting outcome showing that anxiety increases. However, research does show that training away from an attentional bias reduces anxiety [33]. Attentional bias is important in anxiety disorders because attending to threatening information leads to behavioral avoidance of the anxiety provoking and does not allow for exposure to the feared stimuli which would reduce the anxiety in the long term [34].

Several studies have examined whether attentional bias exists in patients with OCD. An early study presented patients with OCD and panic disorder, with OCD threat related words, panic threat related words, general threat related words, and neutral words within a stroop test [35]. There were no significant differences between OCD and panic patients with respect to correct responses on the stroop task. Response latencies also did not differ between panic and OCD participants. Note that longer latencies indicate greater interference or attentional bias. An earlier study presented within the same paper showed no significant differences between the panic and normal comparison groups. Taken together, these results argue against an attentional bias in OCD.

Using a new methodology that involved reaction time to investigate attentional bias, researchers compared people with OCD (predominantly checking behavior) to normal controls and found no significant differences in reaction time to checking related, paranoid theme related, or neutral words [36]. Another study found that an attentional bias was not present in people with OCD compared to normal controls when words were used in a stroop task, but attentional bias was present when pictures were used [37]. In contrast, a series of two studies using the dot-probe task demonstrated that people with OCD exhibited a contamination attentional bias and a general threat attentional bias [38]. However, the general threat attentional bias was equivalent to a high trait anxiety group thereby demonstrating that attentional bias is not specific to OCD. Neither the contamination attentional bias nor the general threat attentional bias was present in a normal control population.

Some evidence for attentional bias in people with an OCD washing compulsion as indicated by longer latencies to name words in a stroop task has been found, although all findings were not supportive of attentional bias [39]. The authors interpreted these results to suggest that the activation of fear in contamination OC washers occurs and interferes with general information processing ability. Some evidence was found for general threat bias in OC nonwashers. Similarly, behavioral treatment of OCD, presumably exposure and response prevention, reduces fear sensitivity as measured by behavioral and physiological measures taken during a cognitive task [40].

Studies have investigated whether attentional bias could be modified in college students with subclinical OCD symptoms [33]. Specifically, the modification intervention focused on drawing participants' attention away from threatening words using a computerized task that displayed threat and neutral words at equal frequencies, but always drew participants' attention to the neutral words by replacing the neutral word with a probe. The modification intervention was administered after establishing that each participant exhibited an attentional bias. The modification intervention group was compared to a control group in which the probe replaced threatening or neutral words at equal frequencies. A behavioral avoidance task using items that participants were led to believe were contaminated served as the outcome measure. Participants who received attentional bias modification training engaged in less behavioral avoidance. They also showed less attentional bias during repeated trials of the attentional bias task administered immediately after bias modification training. An earlier study confirmed the presence of attentional bias in OCD but the bias was limited to the earlier portions of the task [41]. Bias was remediated via habituation in the later portions of the task.

Additional data show that people with OCD exhibit an attentional bias toward threatening information in general as opposed to only OCD related items [42].These data further show that this bias is not due to other aspects of the stimuli that could account for the results, namely emotional arousal [43].

Relatedly, cognitive theories of OCD emphasize cognitive inhibition deficits in OCD, deficits in the allocation of attention via the inhibition of unwanted thoughts [44]. Experimental tests of cognitive inhibition show that people with OCD experience greater difficulty in ignoring previously presented stimuli when instructed to do so relative to people with anxiety disorders other than OCD [45]. Cognitive inhibition becomes further impaired as the complexity of the cognitive task increases [46].

In sum, a substantial amount of evidence exists for attentional bias in OCD. Importantly, attentional bias is not specific to OCD, but rather is present in anxiety disorders in general [47]. Nevertheless, existing studies show that when attentional bias is present in people with OCD, it can be modified, which can reduce behavioral avoidance of distressing stimuli. These data imply that clients should be assessed for attentional bias so that if present, this can be a focus of treatment.

6. Conclusion

6.1. Do the extant data fit the etiological models of OCD?

An examination of the data discussed herein reveals varying levels of support for the components of cognitive models of OCD described at the beginning of this chapter. Consistent support exists for inferential confusion which combines errors in reasoning with devaluation of information taken in by the senses and confusion between real and imagined events. The available data show both higher levels of inferential confusion in people with OCD and reduced inferential confusion following OCD treatment [15[16]. With regard to the role that the tendency to confuse real and imagined events plays in OCD (i.e., reality monitoring), the data are weaker [21,22]. However, it should be noted that reality monitoring ability is the only component of inferential confusion being examined in these studies. However, in inferential confusion processes, reasoning errors and the tendency to discount information taken in by the senses are factors that contribute to the tendency to make reality monitoring errors [14]. Nevertheless, a small amount of data do support the notion that imagery plays a role in the development or maintenance of OCD in people with OCD plus trauma histories [22,24]. However, further research should be conducted in this area. In addition, there is consistent support for the likelihood component of thought-action fusion [8,10,11]. That is, people with

OCD are more apt to believe that having a thought increases the likelihood that the event contained in the thought is going to occur than people without OCD. Less support exists for the belief that having a thought that contains a moral transgression is the equivalent to having completed that transgression (moral component of TAF) [11]. With regard to memory biases in OCD, the data as a whole do not support the existence of memory biases in OCD [27]. However, it is the case that methodological issues in extant studies preclude firm conclusions from being drawn. Moreover, there are numerous studies showing support for the existence of an attentional bias in OCD [36,37,49,41,42]. Also, the data support attentional bias training in reducing OCD symptoms [33]. It should be noted that attentional bias is not a component of the theoretical models specific to OCD described herein, but the data support incorporating attentional bias into models of OCD.

It should be noted that research in this area is in its infancy. Although some but not other components of these models have been tested, extant research has not gone beyond asking the question of whether group differences exist between people with a diagnosis of OCD and various comparison or control groups. Longitudinal studies are needed to test these models holistically. In addition, other components of the models have not been examined. For example, there are no published data on family factors that lend themselves towards vulnerability to creating obsessions. In addition, cognitive appraisals of specific thoughts in people with OCD have not been studied. Moreover, data on the relations between life stress and OCD need to be collected. That is, stress levels in the year preceding the onset of OCD symptoms and overall history of life stress in people with OCD warrant study.

6.2. Treatment implications of cognitive processes and biases in OCD

The theories and data described throughout this chapter indicate that there is substantial room for cognitive components to make a significant contribution to OCD case conceptualization and treatment. Rachman's theory implies that psychoeducation about OCD should begin with presentation of the idea that obsessions are cued by stimuli in the environment [2]. Given that the environment is ripe with cues, clients can expect and should be prepared to expect that the obsessive thought will be cued. If clients are more distressed by obsessions than they are by compulsions, cognitive bias training modification could be employed at this point in the treatment. Modification training would be expected to decrease the obsessions because clients should be less likely to attend to and interpret environmental cues as being related to their obsessions. This would prevent the obsessions from being cued. If cued, clients can be taught that it is the actions that they now take when the obsession is present that are key. Clients should be taught to first, identify when the obsession has come to mind. It is at this point that the client needs to employ mechanisms to cope with the obsession, including cognitive restructuring. The therapist is advised to administer a measure of thought-action fusion such as the Thought-Action Fusion scale [8]. If scores are elevated in the clinical range, the clinician can introduce the idea of thought-action fusion and review each statement endorsed on the TAF as indicating that thoughts equal actions. In order to ward off inferential confusion, clients can be taught inferential reasoning skills that would be expected to reduce the tendency to make reasoning errors. In addition, clients can be taught reality monitoring skills to help remind themselves that a thought is not equal to an action. That is, clients can be taught to ask themselves, "Did I think X thought or did I perform X action?" [adapted from 48]. The therapist can then work with the client to restructure cognitions regarding the probability of events and responsibility for events. Given that the client has likely had a similar thought many times and not acted on it, what is the evidence that they will do so now? With regard to compulsions, clients should be taught about memory norms. First, repeated routine events tend not to be recalled [49]. Therefore, the fact that one cannot remember if one has turned off the stove is normal given the large number of checks performed; therefore, one's memory does not need to be confirmed. Skills from other therapies can be borrowed, such as distress tolerance skills from Dialetical Behavior Therapy [50]. That is, it is OK to feel distress in response to an obsession; one does not need to act on the distress by engaging in compulsions. Distress tolerance skills should be used as opposed to thought suppression skills because research shows that thought suppression can increase rather than decrease unwanted thoughts [51]. Similarly, behavioral activation can be employed so that the client engages in other activities besides rumination. Clients should receive psychoeducation about the normative progression of OCD symptoms. That is, they should expect that new compulsions may occur [52]. They can expect that the OCD will be worse in times of stress and can coincide with negative affect [2]. Consistent with behavioral theory, clients should be taught exposure and response prevention skills; family members that have become part of the OCD rituals should be participants in the therapy [53]. As in many empirically-supported treatments, relapse prevention skills should be taught and clinicians should be sure their clients are able to generalize these skills in order to deal with multiple obsessions and compulsions.

Although inclusion of cognitive components either in addition to exposure and response prevention or as a stand-alone treatment has not been found to add to the effectiveness of OCD treatment, existing treatment outcome studies have not included many of the components discussed in this section [54,55,56] [57]. Therefore, a more comprehensive treatment as proposed may yield more promising results.

Author details

Elisa Krackow*, Sarah Nunley and Pamela Tessier

*Address all correspondence to: elisa.krackow@mail.wvu.edu

Department of Psychology, West Virginia University, Morgantown, USA

References

[1] Abramowitz JS, Taylor S, McKay D. Potentials and limitations of cognitive treatments for obsessive-compulsive disorder. Cognitive Behaviour Therapy, 2005, 34(3) 140-147.

[2] Rachman SS. A cognitive theory of obsessions. Behaviour Research & Therapy, 1997; 35(9) 793.

[3] Julien D, O'Connor KP, Aardema F. Intrusions related to obsessive-compulsive disorder: A question of content or context? Journal of Clinical Psychology, 2009; 65(7) 709-722.

[4] Rassin E, Merckelbach H, Muris P, Schmidt H. The thought-action fusion scale: Further evidence for its reliability and validity. Behaviour Research & Therapy, 2001; 39(5) 537.

[5] Salkovskis PM. Obessional-compulsive problem: A cognitive-behavioural analysis. Behaviour Research & Therapy,1985; 23 571-583.

[6] Beck AT. Cognitive therapy and the emotional disorders. New York: International Universities Press; 1976.

[7] Rachman SS. A cognitive theory of compulsive checking. Behaviour Research & Therapy, 2002; 40(6) 625.

[8] Shafran R, Thordarson MA, Rachman S. Thought-action fusion in obsessive-compulsive disorder. Journal of Anxiety Disorders. 1996; 10 379-391.

[9] Rassin, E., Diepstraten, P., Merckelbach, H., & Muris, P.. Thought-action fusion and thought suppression in obsessive-compulsive disorder. Behaviour Research & Therapy, 2001; 39(7), 757.

[10] Abramowitz JS, Whiteside SS, Lynam DD, Kalsy SS.. Is thought–action fusion specific to obsessive–compulsive disorder?: A mediating role of negative affect. Behaviour Research & Therapy, 2003; 41(9) 1069. doi:10.1016/S0005-7967(02)00243-7

[11] Lee H, Cougle JR, Telch MJ. Thought–action fusion and its relationship to schizotypy and OCD symptoms. Behaviour Research & Therapy, 2005; 43(1) 29-41. doi:10.1016/j.brat.2003.11.002

[12] Barrett PM, Healy LJ. An examination of the cognitive processes involved in childhood obsessive–compulsive disorder. Behaviour Research & Therapy, 2003; 41(3) 285. doi:10.1016/S0005-7967(02)00011-6

[13] O'Connor, K, Aardema F. Fusion or confusion in obsessive-compulsive disorder. Psychological Reports, 2003; 93(1) 227-232.

[14] Aardema, F, O'Connor, K., Emmelkamp, PMG, Marchand, A, & Todorov, C. Inferential confusion in obsessive-compulsive disorder: The inferential confusion questionnaire. Behaviour Research and Therapy, (in press).

[15] Aardema F, O'Connor K, Emmelkamp P. Inferential confusion and obsessive beliefs in obsessive-compulsive disorder. Cognitive Behaviour Therapy, 2006. 35(3) 138-147. doi:10.1080/16506070600621922

[16] Aardema F, Emmelkamp PG, O'Connor KP. Inferential confusion, cognitive change and treatment outcome in Obsessive–Compulsive Disorder. Clinical Psychology & Psychotherapy, 2005; 12(5) 337-345. doi:10.1002/cpp.464

[17] Aardema F, Kleijer TR, Trihey M, O'Connor K. Processes of inference, schizotypal thinking, and obsessive-compulsive behaviour in a normal sample. Psychological Reports, 2006; 99(1) 213-220. doi:10.2466/PR0.99.1.213-220

[18] Johnson M, Raye CL. Reality monitoring. Psychological Review, 1981; 88(1) 67-85. doi: 10.1037/0033-295X.88.1.67

[19] Brown HD, Kosslyn SM, Breiter HC, Baer L, Jenike MA. Can patients with obsessive-complusive disorder discriminate between percepts and mental images? A signal detection analysis. Journal of Abnormal Psychology, 1994; 103 445-454

[20] Krackow E, Rabenshorst M. Does the body know best? Psychophysiological reactivity to perceived versus imagines childhood events. Imagination, Cognition, and Personality. 2010; 30 133-145

[21] Constans JI, Foa EB, Franklin ME, Mathews A. Memory for actual and imagined events in OC checkers. Behaviour Research And Therapy,1995; 33(6) 665-671. doi: 10.1016/0005-7967(94)00095-2

[22] Rubenstein CS, Peynircioglu ZF, Chambless DL, Pigott TA. Memory in sub-clinical obsessive-compulsive checkers. Behaviour Research and Therapy, 1993, 31(8) 759-765. doi:10.1016/0005-7967(93)90006-G

[23] Speckens AE, Hackmann A, Ehlers A, Cuthbert B. Intrusive images and memories of earlier adverse events in patients with obsessive compulsive disorder. Journal of Behavior Therapy and Experimental Psychiatry 2007; 38 411-422.

[24] Rachman SS. Unwanted intrusive images in obsessive compulsive disorders. Journal of Behavior Therapy and Experimental Psychiatry,2007; 38 402-210.

[25] Williams JMG, Watts FN, MacLeod C, Mathews A. Cognitive psychology and emotional disorders. West Sussex, England: John Wiley & Sons Ltd; 1997

[26] Radomsky AS, Rachman SS. Memory bias in obsessive-compulsive disorder (OCD). Behaviour Research & Therapy, 1999; 37(7) 605.

[27] Ceschi G, Van der Linden M, Dunker D, Perroud A, Brédart S. Further exploration memory bias in compulsive washers. Behaviour Research & Therapy, 2003; 41(6) 737. doi:10.1016/S0005-7967(03)00040-8

[28] Wilhelm S, McNally RJ, Baer L, Florin I. Directed forgetting in obsessive-compulsive disorder. Behaviour Research & Therapy, 1996; 34(8) 633.

[29] Radomsky AS, Rachman SS, Hammond D. Memory bias, confidence and responsibility in compulsive checking. Behaviour Research & Therapy, 2001; 39(7) 813.

[30] Tobey AE, Goodman GS. Children's eyewitness memory: Effects of participation and forensic context. Child Abuse & Neglect, 1992; 16(6) 779-796.

[31] Foa EB, Amir N, Gershuny B, Molnar C, Kozak MJ. Implicit and explicit memory in obsessive-compulsive disorder. Journal of Anxiety Disorders, 1997; 11(2) 119-129. doi:10.1016/S0887-6185(97)00001-7

[32] Mathews A, MacLeod C. Cognitive vulnerability to emotional disorders. Annual Review Clinical Psychology, 2005; 1 67-95.

[33] Najmi S, Amir N. The effect of attention training on a behavioral test of contamination fears in individuals with subclinical obsessive-compulsive symptoms. Journal of Abnormal Psychology, 2010; 119(1) 136-142. doi:10.1037/a0017549

[34] Tallis F. The neuropsychology of obsessive-compulsive disorder: A review and consideration of clinical implications. British Journal of Clinical Psychology, 1997; 36(1) 3-20.

[35] Kampman M, Keijsers GJ, Verbraak MM, Näring G, Hoogduin CL. The emotional stroop: a comparison of panic disorder patients, obsessive–compulsive patients, and normal controls, in two experiments. Journal of Anxiety Disorders, 2002; 16(4) 425-441.

[36] Moritz S, Von Mühlenen A. Investigation of an attentional bias for fear-related material in obsessive-compulsive checkers. Depression & Anxiety (1091-4269), 2008; 25(3) 225-229. doi:10.1002/da.20294

[37] Moritz S, Fischer B, Hottenrott B, Kellner M, Fricke S, Randjbar S, & Jelinek L. Words may not be enough! No increased emotional Stroop effect in obsessive–compulsive disorder. Behaviour Research & Therapy, 2008; 46(9), 1101-1104. doi:10.1016/j.brat.2008.05.005

[38] Tata PR, Leibowitz JA, Prunty MJ, Cameron M, Pickering AD. Attentional bias in obsessional compulsive disorder. Behaviour Research and Therapy, 1996; 34(1) 53-60. doi:10.1016/0005-7967(95)00041-U20.

[39] Foa EB, Ilai D, McCarthy PR, Shoyer B, Murdock T. Information processing in obsessive-compulsive disorder. Cognitive Therapy and Research, 1993; 17(2) 173-189. doi: 10.1007/BF01172964

[40] Foa EB, McNally RJ. Sensitivity to feared stimuli in obsessive-compulsives: A dichotic listening analysis. Cognitive Therapy and Research, 1986; 10(4) 477-485. doi: 10.1007/BF01173299

[41] Amir N, Najmi S, Morrison AS. Attenuation of attention bias in obsessive–compulsive disorder. Behaviour Research & Therapy, 2009; 47(2) 153-157. doi:10.1016/j.brat.2008.10.020

[42] Lavy EH, van Oppen P, van den Hout MA. Selective processing of emotional information in obsessive compulsive disorder. Behaviour Research and Therapy, 1994; 32(2) 243-246. doi:10.1016/0005-7967(94)90118-X

[43] Matthew AM, Klug F. Emotionally and interference with color-naming in anxiety. Behaviour Research. & Therapy, 1993; 23 563-69

[44] Beech AR, McManus D, Baylis GC, Tipper SP, Agar K. Individual differences in cognitive process towards an explanation of schizophrenic symptomatology. British Journal of Psychology, 1991; 82 417-426.

[45] Enright SJ, Beech AR. Reduced cognitive inhibition in obsessive-compulsive disorder. British Journal Of Clinical Psychology, 1993a; 32(1), 67-74. doi:10.1111/j.2044-8260.1993.tb01028.x.

[46] Enright SJ, Beech AR. Further evidence of reduced cognitive inhibition in obsessive-compulsive disorder. Personality and Individual Differences, 1993b; 14(3), 387-395. doi:10.1016/0191-8869(93)90307-O.

[47] Mogg, K, Mathews, A, & Eysenck, M. Attentional bias to threat in clinical anxiety states. Cognition & Emotion,1992; 6 149-159.

[48] Giles JW, Gopnik A, Heyman GD. Source monitoring reduces the suggestibility of preschool children. Psychological Science, 2003; 13(3) 288-291. doi: 10.1111/1467-9280.00453

[49] Hudson JA, Nelson K. Repeated encounters of a similar kind: Effects if familiarity on children's autobiographic memory. Cognitive Development, 1986; 1(3) 253-271.

[50] Linehan M. (1993). Skills Training Manual for Treating Borderline Personality Disorder. New York: Guilford; 1993.

[51] Clark DM, Bell S, Pape D. An experimental investigation of thought suppression. Behaviour Research & Therapy,1991; 29(3) 253-257.

[52] Ale CM, Krackow E. Concurrent treatment of early childhood OCD and ODD: A case illustration. Clinical Case Studies, 2011; 10(4) 312-323.

[53] Freeman JB, Garcia AM. Family based treatment for young children with OCD: Therapist guide. New York: Oxford University Press; 2009.

[54] . Abramowitz JS, Taylor S, McKay D. Potentials and limitations of cognitive treatments for obsessive-compulsive disorder. Cognitive Behaviour Therapy, 2005; 34(3), 140-147.

[55] . McLean PD, Whittal ML, Thordarson DS, Taylor S, Söchting I, Koch WJ. Cognitive versus behavior therapy in the group treatment of obsessive-compulsive disorder. Journal of Consulting and Clinical Psychology, 2001; 69(2), 205-214.

[56] Olantunji BO, Rosenfield D, Tart CD, Cottraux J, Powers MB, Smits JAJ. Behavioral versus cognitive treatment of obsessive compulsive disorder: An examination of outcome and mediators of change. Journal of Consulting and Clinical Psychology, 2013; 81(3) 415-428.

http://dx.doi.org/10.5772/58406

[57] Whittal ML, Thordarson DS, McLean PD. Treatment of obsessive-compulsive disorder: Cognitive behavior therapy vs. exposure and response prevention. Behaviour Research and Therapy, 2005; 43 1559-1576.

Chapter 3

The Utility of Electromagnetic Activity Measures in Obsessive Compulsive Disorder and Schizophrenia

Elana Harris and Milena Korostenskaja

Additional information is available at the end of the chapter

http://dx.doi.org/10.5772/57247

1. Introduction

A young adult presents to a psychiatrist and describes the need to "unwind" himself around pieces of furniture if he has passed them in a clockwise direction. When he is weighed, he expresses the concern that pieces of rubber will come off the office scale and that will be considered stealing. He hesitates before he sits down on a chair in the office because he is afraid that he will catch the emotional problems of the person who sat in the chair before him. Are these psychotic delusions or obsessions and compulsions?

A delusion is a fixed, false, unshakable belief. An obsession is a worried thought or image that is experienced as disturbing. The distinction is important to make as the treatment paths for obsessive compulsive disorder (OCD) and schizophrenia (SCZ) diverge.

The aim of this chapter is to review electromagnetic measures that may be used to help with diagnostic clarification and may predict treatment response in these two psychiatric disorders. At present, electroencephalograms (EEGs) or other neurophysiological measures are not routinely ordered for psychiatric patients. They have not yet become the standard of care. The Food and Drug Administration (FDA) recently approved the Neuropsychiatric EEG-Based Assessment Aid (NEBA) System to assist in the diagnosis of attention deficit hyperactivity disorder (ADHD) based on the ratio of theta to beta frequency bandwidths. Electrophysiological tests to aid in the diagnosis and management of patients with OCD and SCZ may develop in the near future.

2. Similarities and differences between OCD and SCZ: Epidemiology and clinical signs

2.1. Epidemiology

The prevalence of both OCD and SCZ is high, with estimates across the globe ranging from 1-2% of the population [1-4]. There is significant morbidity associated with both of these conditions [5]. Symptoms of these diseases may impact the individual's family, and ability to work and contribute to society. The quality of life of patients with OCD is impaired primarily during the symptomatic state but less so when patients are treated or in remission [6, 7]. Close to 76% of patients with SCZ are unable to engage in basic social roles, even when psychotic symptoms are in remission; few marry, and less than one third are in regular employment [8]. Nine to thirteen percent of patients with schizophrenia commit suicide [9].

The prevalence of obsessive compulsive symptoms in patients with SCZ ranges from 7.8% [10] to 25% [11]. Whereas less than 2% of patients with OCD develop psychotic symptoms [12]. The age of onset of OCD is bimodal with peaks both in children and young adults [13]; the typical age of onset of SCZ is in the third decade of life with childhood onset being extremely rare [14].

2.2. Clinical characteristics

The clinical presentation of OCD and SCZ is currently what is used to diagnose and distinguish these conditions. OCD is characterized by the presence of obsessions and/or compulsions [15]. Obsessions are unwanted thoughts or images that recur. Compulsions are repetitive behaviors or mental acts that an individual feels driven to perform. Patients with OCD often describe either a sense of incompleteness if a ritual is not done just right or a sense that something bad will happen if they don't perform the ritual. SCZ, on the other hand, is characterized by positive and negative symptoms [15]. Positive symptoms refer to additional symptoms that are not present in a healthy individual such as hallucinations, delusions, and thought disorder. Negative symptoms refer to deficits, such as, lack of facial expressions and emotional variability, decreased energy and diminished verbal output. Cognitive dysfunction [16] and disorganized behavior may be present as well and include disorganized speech, bizarre behavior and poor attention [17].

While OCD and SCZ are described as distinct psychiatric disorders [18], some authors argue that a "schizo-obsessive disorder" exists as well [19-21]. Indeed, a subset of individuals with SCZ present with obsessive compulsive symptoms and a subset of patients with OCD lack insight. Some have concluded from this overlap, that there is a spectrum of disorders that ranges from: [1] OCD, [2] OCD with poor insight, [3] OCD with schizotypal personality disorder, [4] schizophrenia with obsessive compulsive symptoms, [5] SCZ with OCD and [6] SCZ [22].

http://dx.doi.org/10.5772/57247

3. Electromagnetic measures of neural activity

3.1. Transcranial magnetic stimulation (TMS)

TMS is a versatile tool in the hands of a neurophysiologist as it can be used to measure and modulate cortical excitation and inhibition. The TMS unit consists of a strong (1 to 2 Tesla) electromagnetic generator and a handheld magnet or adjustable coil. When stimulating, the coil is positioned manually over the scalp. Some systems also include a means by which the investigator may navigate and visualize the location of stimulations by co-registering the head position with a 3-dimensional reconstruction of the subject's own MRI. The magnet can be set to deliver single or repetitive pulses generating focal electrical currents.

With the magnet held over the contralateral primary motor cortex, a single magnetic pulse excites the underlying brain tissue and leads to an evoked potential and movement in the corresponding muscle. The amplitude of the motor evoked potential (MEP) and the motor threshold (or level at which 50% of the stimuli lead to movement) reflects the degree of excitability of the brain, spinal cord neurons and muscles in that individual. A period of inhibition, typically lasting for a few hundred milliseconds, follows the MEP. This cortical silent period (CSP) is obtained by asking the subject to maintain muscle contraction while a single suprathreshold TMS pulse is applied to the motor cortex [23-25].

Neuroplasticity is a feature of the nervous system that helps the brain learn, develop or reorganize in response to intrinsic or environmental stimuli. In broad terms, though such reorganization can be associated with the development of a healthy skill or recovery after a functional loss such as a stroke, maladaptive changes may lead to problematic patterns of thoughts and behaviors. The underlying mechanism behind the strengthening or weakening of neuronal connections is supported by in vivo and in vitro animal experimentation and is thought to be based upon long term potentiation (LTP) or long-term depression respectively (LTD) [26, 27]. More recently, several TMS protocols have been developed to study the inhibition and facilitation of MEPs which may reflect the underlying influences of inhibitory and excitatory cortico-cortial and subcortico-cortical circuits which modulate cortical excitability.

Paired-pulse TMS is a method of applying stimuli below the MEP threshold to change the size of subsequent MEP. A single "conditioning" pulse is followed by a "test" pulse. The interstimulus interval (ISI) affects the size of the resultant MEP. In short interval intracortical inhibition (SICI) protocols, a subthreshold stimulus is followed by a suprathreshold stimulus. Interstimulus intervals of 1-5ms lead to suppression of the MEP. With long interval intracortical inhibition (LICI) both pulses are suprathreshold, and the interstimulus interval is 50-200ms. MEPs can be facilitated when a subthreshold pulse is given 10-25ms before a suprathreshold pulse. Research suggests that intracortical inhibition and facilitation reflect the influences of inhibitory and excitatory cortio-cortical and subcortical–cortical circuits modulating activity in motor cortex output neurons without the involvement of spinal neurons [28-30].

More durable changes to cortical excitability which persist past the stimulation period can be induced utilizing various repetitive TMS (rTMS) protocols. Already in wide use for the treatment of depression, 1 Hz rTMS can also be used experimentally to temporarily disrupt cortical activity and thus establish structure-function relationships when used in tandem with behavioral experiments and functional imaging. Higher frequency rTMS (5 to 20 Hz) tends to increase cortical excitability. Although 1-Hz rTMS (applied at 90% of resting motor threshold) to the contralateral motor cortex for 10 minutes results in approximately 10 minutes of MEP size depression following stimulation, recent protocols that utilize more complicated patterns of stimulation can result in effects that last longer than 30 minutes. Theta-burst stimulation, for example, consists of ultra-rapid trains of three TMS pulses (50 Hz) with variable interval between bursts for a total stimulation time of 40 to 190 seconds. Remarkably, this brief stimulation can result in relatively long-lasting changes in MEP size resembling a human model of LTP and LTD of synaptic efficiency [31].

TMS has an important role in research regarding the nervous system and its role in the treatment of various psychiatric conditions is expanding (for review, see 32). It has been shown to be useful in reducing auditory hallucinations [33, 34] and improving negative symptomatology in SCZ [35]. The jury is still out on whether rTMS will prove useful for improving symptoms in OCD; the three studies to date do not support significant benefit of rTMS [36].

TMS studies in OCD. Just as there are few treatment studies utilizing TMS in OCD, so too, few investigator have employed TMS to examine cortical excitability in OCD. The first study we found in the literature, reported decreased SICI in 12 patients with OCD compared with 12 healthy comparison subjects (HCS) [37]. This group expanded upon their findings using both single-pulse TMS and paired-pulse TMS in 9 medicated and 7 unmedicated patients with OCD compared with 11 HCS [38]. They found a lower motor threshold both when the OCD subject was stimulated at rest or during an active state. They also found diminished SICI in the OCD subjects, with even lower intracortical inhibition in subjects with comorbid OCD and a tic disorder. There were no significant differences between subjects with OCD and HCS with regard to MEP amplitude, intracortical facilitation or length of the silent period. In a larger sample, Richter, de Jesus [39] compared 34 patients with OCD (23 medicated and 11 unmedicated) with 34 HCS. In contrast to the previous study, no difference was found in resting motor threshold between the OCD and HCS, although the resting motor threshold was significantly lower in the OCD subjects on medication. The CSP was shorter in patients with OCD compared with HCS. No differences were found in SICI between the OCD and HCS, but patients with OCD had greater intracortical facilitation. No correlations were found between illness severity and TMS parameters in either the medicated or unmedicated patients. The discrepancies between these studies may reflect the presence of unmedicated subjects or may be attributed to different TMS stimulus parameters. However given the paucity of studies of TMS in subjects with OCD there is a need to continue research to further our understanding of the possible excitation inhibition imbalance in this disorder.

To explore the mechanism of action of rTMS in subjects with OCD, recently Pedapati, DiFranscesco [40] examined the effects of 30 minutes of 1 Hz repetitive TMS (rTMS) of the dorsolateral prefrontal cortex. We compared sham (subthreshold) TMS with rTMS on the blood oxygenation level-dependent (BOLD) signal during symptom provocation and found increased BOLD activity in the right inferior frontal gyrus, right insular cortex, and the left thalamus in the sham subjects suggesting that rTMS may have inhibited the desensitization process experienced by the subjects during provocative image exposure (Figure 1).

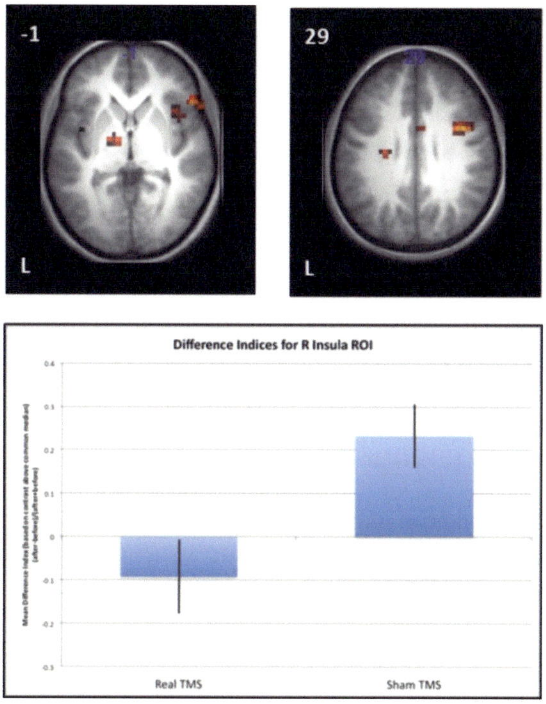

Figure 1. Differences in brain activation for subject-specifc OCD symptom-provocative task comparing 1 Hz rTMS with sham rTMS. Top: Select axial slices show an interaction between the intervention (real or sham rTMS) and time (pre- or post-rTMS). Hot colors indicate (sham/after > sham/before) > (TMS/after > TMS/before) contrasts. Colors indicate activations that passed an uncorrected threshold of $p < 0.005$. Neurological convention used. Bottom: Results of regional analysis for a region of interest covering the right insula. A difference index comparing activation before and after rTMS is shown for the real and sham rTMS subject groups. The group difference is significant at $p<0.05$. Black bars indicate the standard error.

TMS studies in SCZ. There is a somewhat larger body of literature examining the effects of TMS in SCZ that is less divergent in its findings than the OCD literature. Abnormalities in cortical inhibition in patients with SCZ have been reported by a number of authors [41-43]. Eichhammer, Wiegand [44] found that treatment naïve patients with SCZ had significantly lower resting motor threshold relative to healthy subjects. Liu, Fitzgerald [45] found that patients with SCZ who were treated with clozapine had longer CSP compared with other patients with SCZ, while Wobrock, Schneider-Axmann [46] found prolonged CSP in patients with new onset SCZ who had limited exposure to medication. Daskalakis, Christensen [47] found deficits in use-dependent plasticity in subjects with SCZ which is measured after subjects have been trained to move in the opposite direction of the movement that is induced by TMS.

Reduced SICI has been recorded in first-episode SCZ and has been found to correlate with positive symptom severity [46, 48]. Some medications have been found to affect TMS measures of cortical inhibition. For instance, Fitzgerald, Brown [49] found that olanzapine and risperidone confer different effects on the resting motor threshold and cortical inhibition. It may take time or a dose effect to notice medication changes as Daskalakis, Christensen [50] have pointed out that single doses of haloperidol and olanzapine did not alter cortical inhibition in healthy subjects.

In summary, relative to HCS, patients with OCD were found, by at least some investigators, to have lower motor threshold, shorter CSP, decreased SICI, and greater intracortical facilitation. Patients with SCZ were also found to have lower motor threshold and decreased SICI, but prolonged CSP and abnormalities in use-dependent plasticity. Thus the only measure to date which may distinguish between these conditions to date, using TMS, is the length of the cortical silent period. For the development of a useful diagnostic measure, head-to-head studies are needed for direct comparison between these and other psychiatric conditions. No studies to our knowledge have used TMS measures to predict treatment outcome, although, as discussed above, TMS measures can detect changes that result from treatment with antipsychotics.

3.2. Electroencephalography (EEG) and magnetoencephalography (MEG)

The neural origins of brain function in psychiatric patients can also be effectively studied with non-invasive neurophysiological techniques such as electroencephalography (EEG) and magnetoencephalography (MEG) [51-53]. Handy (2009) notes that for a time investigators began to consider EEG to be a tool of the past, but functional magnetic resonance imaging helped revive an interest in combining the complimentary anatomical and electrophysiological approaches and there is now an upsurge of interest in EEG and MEG. Both EEG and MEG can be recorded in patients at rest with eyes open or closed (spontaneous EEG or MEG) as well as during cognitive or behavioral tasks. Signal analysis techniques allow for quantitative interpretation of both EEG and MEG waveforms – qEEG/qMEG, respectively. Such analysis can be helpful not only in the diagnosis of psychiatric conditions [54], but also in predicting treatment outcome [55]. Moreover, recent development in functional connectivity analysis, permit investigators to study the activity in disparate brain regions in psychiatric patients at rest with MEG or EEG. This approach has been referred to as the study of "resting state functional connectivity" or the "default mode network." [56]. Indeed, since there is evidence to suggest that the core feature of disorders like OCD and SCZ are a result of altered functional connections between different brain regions [57, 58] this approach is likely to prove to be very valuable.

EEG and MEG add, to the already rich functional imaging literature, the ability to record neural activity with high temporal resolution. EEG uses surface scalp electrodes to monitor cortical electrical potentials. Electrodes distributed across the scalp together with mathematical analyses can estimate the location of the generator of the neural activity within several centimeters. Measurements with MEG permit 3-D localization of current sources studied on a time scale of less than 1 ms [59]. MEG uses magnetometers to record the magnetic fields

http://dx.doi.org/10.5772/57247

produced by extracellular electrical currents. The most common magnetemotors in use are referred to as superconducting quantum interference devices or SQUIDS. Subjects are seated or supine during recordings with a helmet containing the SQUIDS placed over their heads. Cortical activity is on the order of 10 femtotesla (fT) and the alpha rhythm runs on the order of1000 fT. These magnetic fields are much smaller than the background noise, which is on the order of 10^8 fT, thus various strategies are employed to remove the noise, including magnetically shielding the recording room. With three fiducial markers (typically at points on the nose and ears), MEG data can be aligned with the subject's own anatomical magnetic resonance image (MRI). MEG and MRI data can further be transformed to Tailarach space to assist in group comparisons.

One of the principal differences between MEG and EEG is thought to be that MEG mainly records activity from tangentially oriented sources leading to better recordings from sulci where the pyramidal cell dendrites are lined up parallel with the cortex, while EEG can detect sources with both radial and tangential components resulting from intracellular currents [51, 60, 61]. If one uses a perfect sphere to model the brain, there would be no magnetic field detected from an entirely radially oriented current dipole. However, the human brain has a more complex shape and most current dipoles have both radial and tangential components. Therefore, a nearly radial source in the brain may generate a magnetic field that can be detected by MEG [62]. In both mathematical modeling studies and in animal experiments, the strength of the sources has been found to differ based on the location and orientation [63-65]. Sensitivity to tangential sources makes MEG highly relevant for studies of auditory processing. For example, MEG can be particularly sensitive to studying auditory hallucinations in SCZ or auditory sensory information processing in OCD, as temporal brain regions are thought to be involved in pathophysiology of both of these disorders [66-68]. A further difference between EEG and MEG recordings is that magnetic fields are less distorted than electric fields by the skull and scalp, leading to better spatial resolution for MEG. Thus, although EEG is sensitive to activity both at the tops of the gyri and in the sulci, activity that is recorded with MEG can be localized better.

Since frontal regions play a prominent role in the pathophysiology of both OCD and SCZ, one might wonder how effective MEG is at detecting neuronal activity in the orbitofrontal cortex (OFC), for instance. Hillebrand et al. [69] addressed this issue with detailed computations of MRI modeled brain gyral surfaces, realistic strength cortical sources, and realistic background noise. As a result of the study, the authors found limitations for MEG sensitivity only for the most posterior aspect of the OFC. A careful study comparing MEG and fMRI localization of responses to emotionally laden pictures showed co-localization of MEG and fMRI activation of orbitofrontal cortex within 7-9 mm [70]. More recently, others [71, 72] have confirmed MEG sensitivity to source activity in the OFC.

In quantitative electroencephalography (QEEG), multichannel recordings, usually from the standard 19 electrode positions, are obtained while the subject has his or her eyes-closed and is in a relaxed, awake state. One to two minutes of artifact-free data is analyzed using the Fast Fourier Transform to quantify the power at each frequency of the EEG averaged across the recording time. This is referred to as the power spectrum. There is very good test-retest reliability of power spectra computed in this manner. Power spectra are typically examined in the range of 1 to 20 Hz frequency band. This frequency range is further divided into frequency bands and assigned names from the Greek alphabet: delta (δ, 1.5—3.5 Hz), theta (θ, 3.5—7.5 Hz), alpha (α, 7.5—12.5 Hz), and beta (β, 12.5—20 Hz). Results of the analyses describe absolute power in each frequency band, the relative power or percentage of total power in each channel, coherence which measures synchronization between two channels, or symmetry which is the ratio of power in each band between a symmetrical pair of electrodes [73].

EEG/MEG in OCD. The first known EEG study regarding patients with OCD patients reported seizure like patterns with slow waves of 2–4 Hz [74]. Similarly, nonspecific theta activity was found in the EEG of patients with OCD [75]. To date, the δ, θ, α and β frequency bandwidths have been examined in patients with OCD in the awake, resting state and during the performance of cognitive tasks. Table 1 illustrates the large variability in findings between different investigators. Both increased α and β [76-79], as well as, decreased α and β have been found in patients with OCD [80-85]. There are also reports of lateralized left [86] and right hemispheric differences in patients with OCD [87, 88]. What the majority of the studies have in common is a deviation from the healthy comparison group in frontal and frontotemporal regions [78, 79, 81, 83, 84, 87, 89].

Method	Population	Results for OCD patients (pts)	Authors, year
Resting EEG intracortical exact low resolution electromagnetic tomography software	30 drug-free pts with OCD and 30 HCS	↓ lagged non-linear coherencee β2 frequency between frontal brain areas but not within the default mode network. High vigilance stages had yielded ↓ frontal phase synchronisation for β and θ	Olbrich, Olbrich [90]
Resting EEG with standardized low-resolution electromagnetic tomography software	50 OCD (8 drug-free; 42 on SSRIs), 50 HCS	↑ δ in the cingulate gyrus, did not correlate with symptom severity or illness duration. δ power in the right orbitofrontal cortex positively correlated with age of OCD onset	Koprivova, Horacek [91]
Resting EEG with low-resolution electromagnetic tomography and coherence analysis	37 drug naïve pts with OCD and 37 HCS	↑ δ in the insula ↑ β in frontal, parietal and limbic regions. Decreased interhemisphereic coherence Reduced coupling between δ and β	Velikova et al. [79]
Quantitative EEG	20 adults (10 M, 10 F) with OCD and 19 HC	↑ in θ coherence in the fronto-occipital region	Desarkar et al. [92]
Resting EEG with variable resolution electromagnetic	20 adults with OCD treated with paroxetine	↑ α in the striatum, orbito-frontal and temporo-frontal regions pre-treatment. This abnormality decreased following successful treatment with paroxetine.	Bolwig et al. [76]

http://dx.doi.org/10.5772/57247

Method	Population	Results for OCD patients (pts)	Authors, year
tomography			
Quantitative EEG	18 unmedicated adults with OCD and 18 HC	Lower average background frequency in frontal regions. ↑ δ - and ↓ α / β power.	Pogarell et al. [81]
Multilead QEEG while subjects performed executive function tasks	32 drug-free pts with OCD and 32 HC	↓ α power	Bucci et al. [80]
Rey-Osterrieth Complex Figure Test	23 pts with OCD	Score correlated with the α power with regression coefficients that had different directions by hemisphere	Shin et al. [93]
neurometric QEEG	20 nondepressed pts with OCD all treated with paroxetine	↑ α at baseline in medication responders.	Hansen et al. [77]
19-channel QEEG recorded at rest in supine subjects	32 drug-free patients (8 males) with OCD and 31 HCS (9 males)	↓ absolute β power and an ↑ relative θ power in frontotemporal regions There were also differences by OCD subtype	Karadag et al. [84]
QEEG during resting state and during hyperventilation	22 unmedicated nondepressed pts with OCD and 20 HC	At rest: ↑ δ & θ; ↓ α in left frontotemporal regions During hyperventilation: ↓ β in left frontal regions	Tot et al. [82]
self-paced movement of the right thumb; 29-channel EEG	10 untreated OCD patients and 10 HC	Delayed onset of mu event-related desynchronization with movement preparation and less postmovement β synchronization	Leocani et al. [94]
QEEG during live and imaginal exposure to feared contaminants	6 pts with OCD	significant change in the anterior-to-posterior scalp distribution of α power during live exposure	Simpson et al. [96]
rest and during a temporal lobe activating procedure, i.e., olfactory stimulation	37 drug-free patients with OCD and 30 HCS	At rest: ↑ delta-1 and ↓ α -2 power. During olfactory stimulation: HCS had a power increase in β, whereas OCD patients showed no change (right) or slight decrease (left)	Locatelli et al. [83]
EEG spectral analysis	50 pts with OCD, 50 pts with anxious neurosis, 25 HC	↑ mean α power in occipital regions. ↓ frontal β activity found in both patient groups	Serra et al. [85]
QEEG	27 adult patients with OCD in a drug-free state at baseline	Found 2 groups: Cluster 1 - ↑ relative power θ in the frontal and frontotemporal regions; 80% of these patients did not respond to medication. Cluster 2 - ↑relative power α, 82.4% of these patients were treatment responders.	Prichep et al. [78]

Table 1. Summary of EEG findings in OCD

Spontaneous MEG in OCD. In the first documented MEG study with OCD subjects, the findings of Amo, Quesney [71] are reminiscent of the early EEG studies in OCD (for example, 74) with fronto-temporal paroxysmal rhythmic activity with low-amplitude spikes and intermittent isolated spikes and sharp waves. Amo, Quesney [71] expanded their findings with the addition of two selective serotonin specific reuptake inhibitor-naïve subjects [97] and they found similar paroxysmal activity in fronto-temporal regions. Maihofner, Sperling [98] observed the spontaneous MEG activity in 10 subjects with OCD and 10 healthy subjects. These authors parsed the frequency bandwidths examined into "slow" (2-6 Hz) and "fast" (12.5-30 Hz). They found dipole maxima concentrated in the left superior temporal gyrus with no difference in the number of dipoles between subject groups for slow MEG activity. However, the OCD group had a clustering of slow MEG activity over the left dorsolateral prefrontal cortex.

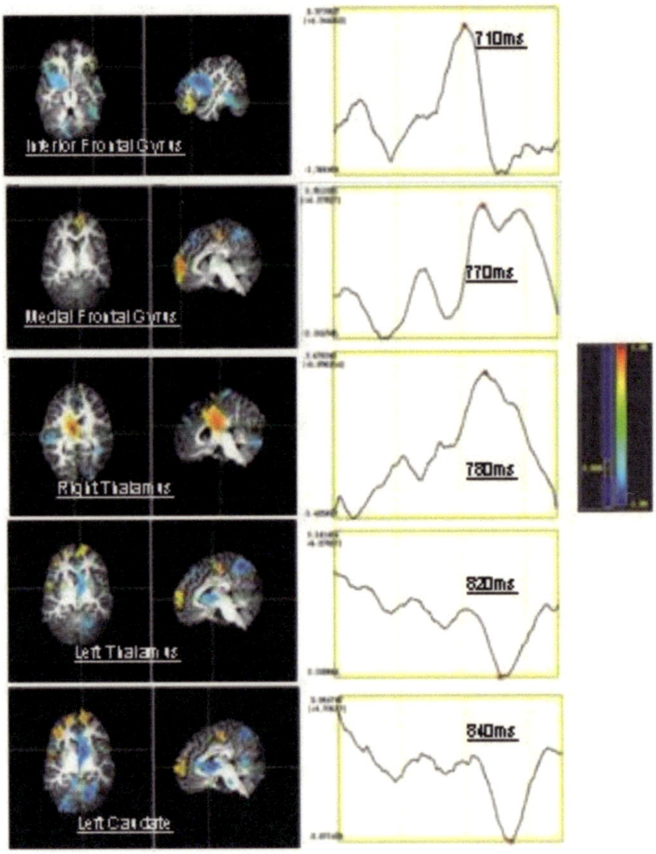

Figure 2. Example of MEG data in an Adolescent with OCD following the presentation of Symptom Provocative Images. Event-related beta synchronization (warm colors) and de-synchronization (cool colors) are shown in frontal cortical regions, thalamus and caudate in an adolescent subject with OCD who was presented with contamination related stimuli. Left to right, Axial, parasagittal views (Left=Right), and graph of time course of activity following stimulus presentation. The red circle in the graph indicates the time at which the peak activity occurs within the green cross-hairs in the corresponding MRI.

http://dx.doi.org/10.5772/57247

*Task-related MEG in OCD.*To explore the network and compensatory mechanisms, Ciesielski, Hamalainen [99] examined the MEG pattern of activation during a working memory task in subjects with OCD. They found that during the encoding phase, there was enhanced activity in OCD subjects in the anterior insula and decreased activity in the posterior inferior parietal cortex. During the retention phase, activity was lower in the occipital, parietal, superior temporal sulcus and dorsolateral prefrontal cortex. During the retrieval phase there was a significant increase in activation from the right anterior insula extending toward the orbital region and the right superior temporal sulcus. There was reduced activity in the left parietal cortex. Ciesielski et al. [100] examined event-related synchronization (ERS) and desynchronization (ERD) in the alpha band associated with a working memory task in subjects with OCD and healthy subjects. In subjects with OCD, these authors found lower baseline alpha and a task phase-specific ERD.

In a small sample of adolescents with OCD, we have begun to look at the power of oscillations at rest and during a symptom provocative task [101, 102]. The majority of images for the visual task were the neutral and contamination sets of images previously used by Gilbert, Akkal [103] The spatiotemporal structure of beta band event-related changes were analyzed with synthetic aperture magnetometry together with a Stockwell transform to provide power as a function of time for each voxel. Results from one individual are illustrated in Figure 2 demonstrating the ability of MEG to show the precise timing of activity of elements within the circuit.

With regards to functional connectivity analysis of electromagnetic data, to our knowledge, only one study in OCD has been conducted to date [90]. When recording resting state EEG, these authors demonstrated altered functional connectivity within the frontal brain region (decreased non-linear coherence within the beta-2 frequency band) in OCD group compared with HCS.

Overall, although the electrophysiological literature is sparse, both EEG and MEG data support other functional imaging modalities (Beucke et al. 2013; [104] in their implication of elements of a frontocortical, striatal, thalamic circuit with involvement of limbic regions, in line with current neural model of OCD [105].

Spontaneous EEG/MEG in SCZ. A large number of studies have demonstrated that EEG abnormalities occur more frequently in patients with SCZ than in healthy subjects. As early as 1936, Lemere [106] wrote "... the apathy and affective deficiency of the schizophrenic was the feature of the illness most clearly related to an absent or 'poor' alpha rhythm." Berger, too, in 1937 [107], noted alpha and beta frequency abnormalities in a patient with schizophrenia. Early reports also suggested that there were statistically significant resting EEG differences between healthy individuals and patients with SCZ [108-116]. Abnormalities include general slowing, dysrhythmia, nonspecific diffuse patterns, atypical sharp waves and epileptiform discharges. A discussion in the literature ensued regarding the relationship between schizophrenia and epilepsy. In support of this connection were the slow waves and spikes that were recorded during catatonic episodes [112, 117, 118]. Still in all, a number of early studies failed to find any EEG abnormalities in patients with SCZ, e.g., Colony and Willis [119] did not find EEG differences between their 1000 patients with SCZ and HCS. Abnormal EEG activity was thought of by some to be the result of a premorbid head injury or the consequence of treatment with electric, insulin or chemical shock [120-122]. Sponheim et al. [123, 124] looked for correlations to explain the variance in central EEG slow wave prevalence. Patterns found include a relationship that favored winter birth over diagnosis. Patients with more negative symptoms and larger ventricles had an increased likelihood of slow wave abnormalities. In a landmark study, Shagass et al. [125] compared the EEGs of patients with SCZ with patients with other psychiatric conditions and HCS. They reported a sensitivity of 50% and specificity of 90% when patients with SCZ were compared to patients with Major Depression. Their work was replicated by Gerez and Tello [126] who used a battery of 10 neurophysiological assessments and found 78% sensitivity and 85% specificity in classifying subjects. However, Sponheim et al. [127] were unable to differentiate patients with SCZ from patients with affective disorders using low frequency and alpha band EEG power, but they succeeded in using these measures to differentiate SCZ from healthy subjects.

The intrinsic EEG of SCZ show augmented theta/delta and reduced alpha power (for example, 128, 129, 130). These abnormalities correlate with psychotic symptoms [131, 132], candidate risk genotypes [129] and perithalamic ventricular volume [123]. Stimulus elicited low frequency (delta-alpha) phase locking and single trial power is also consistently reduced in SCZ [133, 134], an effect that is highly heritable [135]. Importantly, thalamic aberrations have been theorized to be relevant both for SCZ neuropathology and the expression of psychotic symptoms [136, 137]. Klimesch, Sauseng [138] suggest that the heritability and consistency across paradigms with regard to alpha/theta/delta oscillations should be a considered an EEG marker of thalamocortical disconnectivity in SCZ.

Siekmeier and Stufflebeam [139] reviewed the resting state MEG literature for patients with SCZ from 1993 to 2009 and found that there was overwhelming support (11/12 studies) for increased theta (4-8 Hz) and delta (1-4 Hz) band oscillations in the temporal lobes of patients. Of the studies that correlated oscillations with symptoms, there was a positive correlation (in 8/10 studies) between temporal lobe theta activity and positive symptoms.

As Boutros, Arfken [140] point out in their meta analysis that included 15 studies of spontaneous EEG comparing subjects with SCZ to healthy subjects or non-schizophrenic psychotic patients, a large number of statistically significant differences have been found. However, they go on to note, no systematic effort using a large multicenter population has been made to standardize an assessment battery. Further research in this area is warranted.

Task-related EEG in SCZ. In addition to resting state analyses, EEG oscillations can be examined during specific phases of cognitive tasks. For example, Dias, Bickel [141] compared the responses of patients with SCZ and HCS during the "AX" continuous performance task. In this task, subjects are asked to attend to a sequence of individually presented letters and must respond whenever they view a letter "A" followed by a letter "X," and ignore all other sequences. They found task-related event-related desynchronization that was reduced in the beta band in the parieto-occipital cortex for sensory encoding in and reduced beta ERD in the motor cortex during response preparation in patients with SCZ.

Gamma EEG oscillations in humans can be stimulated by a task, induced by a stimulus, or evoked by repetitive inputs. In almost all cases, the amplitude of gamma is reduced in SCZ

[142]. Gamma oscillations are thought to support the cognitive processes (e.g., attention, memory, and learning) that are disrupted in SCZ, and these oscillations are thought to facilitate communication between brain regions involved in SCZ. This has given rise to the hypothesis that abnormalities in gamma oscillations may be one of the principal underlying problem in patients with SCZ [143]. In support of this notion, many studies have found that patients with SCZ to have decreased power or synchrony of gamma oscillations during responses to sensory stimulation or cognitive tasks [144-149]. In some cases, these abnormalities correlate with the severity of cognitive dysfunction or other symptoms [145, 148]. Despite the decreased power or synchrony of gamma oscillations evoked by sensory stimuli or cognitive tasks, surprisingly, auditory hallucinations are apparently linked with increased power or synchrony of beta and gamma oscillations [150-152].

Functional connectivity studies in SCZ. As highlighted above by the opposing valence of the changes in gamma oscillations in SCZ dependent on the task or setting, examination of a single frequency range in a single brain region may yield insufficient information to be a reliable biomarker of disease state. Bassett has suggested that functional connectivity disturbances would make excellent diagnostic biomarkers for the disease [153]. The examination of oscillations across brain regions using network theoretical tools borrowed from the social sciences indeed provides support for "dysconnectivity" in patients with SCZ [153-165]. Allen, Liu [166] demonstrated that cross-frequency interactions are abnormal and increased or decreased in various regions in SCZ, with the strength of these interactions correlating with genetic risk factors for the disease. Hinkley, Vinogradov [58] too performed resting-state functional connectivity analysis of MEG data (alpha frequency band) with eyes closed in SCZ patients and compared it with HCS. In SCZ patients, left prefrontal cortex as well as right superior temporal cortex had decreased connectivity; at the same time functional connectivity in left extrastriate cortex and in the right inferior prefrontal cortex were increased. Importantly, these latter changes in the right inferior prefrontal cortex correlated with cognitive deficits in SCZ patients. Important results were demonstrated by Higashima, Takeda [167], who by using functional connectivity analysis approach of resting state EEG data, showed that there is a functional disconnection between left and right frontal lobes in schizophrenia patients and with normalization following antipsychotic treatment.

Most recently, Siebenhuhner, Weiss [154] examined the functional connectivity in 14 patients with SCZ and 14 HCS using MEG during a working memory N-back task. Their analysis was based on a multiresolution approach [159] which posits that neurophysiological alterations in SCZ manifest as a complex hierarchy of signatures across univariate (individual sensor time series or entropy), bivariate (co-variability between time series or functional connectivity), and multivariate (patterns of co-variability across sensors or network topology) statistical measurements. In addition, they examined functional networks constructed from the interactions between frequency bands. They found an extensive pattern of altered network structure and network dynamics in patients with SCZ with disease-associated changes in brain function at each level of analysis. Patients with SCZ had lower time series entropy and increased strength of co-variability between time series. These findings were suggestive of decreased information content of MEG signals and, perhaps surprisingly, hyperconnectivity between brain regions.

They also found that patients with SCZ had deviant topological organization in binary sensor networks and that network properties of cross-frequency associations between time series in the beta and gamma bands differed between groups.

Overall, there is a strong potential of functional connectivity analysis to contribute to diagnosis and treatment in SCZ since the essential feature of this disorder is thought to be one of functional disconnection between brain regions.

In our review of the literature, we did not find any comparative studies that examined groups of subjects with OCD and directly compared them with subjects with SCZ using spontaneous EEG or MEG recordings or functional connectivity analyses. This work would be valuable to aid in the development of diagnostic tests that differentiate between these conditions.

3.3. Information processing studied with event related potentials and fields (ERPs/ERFs)

Signal averaging helps extract event-related potentials (ERPs, recorded with EEG) and event-related fields (ERFs, recorded with MEG), brain responses specifically related to the external or internal stimuli, from background spontaneous brain activity. ERPs/ERFs studies contributed significantly to understanding neural basis of both OCD and SCZ, and specifically the neural origins of cognitive dysfunction of these disorders [168]. Moreover, some of ERPs are proposed to be used as biomarkers of each of these specific disorders [169, 170]. For a summary of information about ERP studies in OCD – see Table 2). The main ERP/ERF responses studied in OCD and SCZ to date are related to early sensory processing, attention and performance monitoring.

Information processing at the brain stem level. In OCD, it has been possible to demonstrate changes in processing of auditory information stream already at the level of brain stem [85, 171]. For instance, Nolfe et al. [171] by recording brain stem auditory evoked potentials (BAEPs) showed that the interpeak latency of wave I-V was delayed, as well as amplitude of wave III was reduced in OCD patients when compared with HCS.

At the same time, BAEPs results observed in SCZ are variable [172, 173], with some of them including modified BAEPs among patients with positive symptoms [173]. More recently, Kallstrand, Nehlstedt [174] has demonstrated significant interhemispheric differences in wave II of BAEPs in SCZ patients when compares with HCS, which may imply that lateralized abnormalities exist already on the initial level of auditory information processing in the brain.

Overall, information processing alterations at the brain stem level seem to be better documented and more consistent in OCD patients rather than in patients with SCZ. Comparative studies are needed to evaluate differences on the initial stages of auditory information processing in the brain in these two patient groups.

Sustained neuronal entrainment to repetitive stimulation. A remarkable opportunity to study how brain activity is synchronized with the external events is provided by steady-state evoked responses (SSR) (for review, see 175]. SSRs can be evoked by the trains of repetitive auditory (ASSR) [176], visual (VSSR) [177] or multi-sensory (audio-visual) [178] stimuli. At certain frequencies, these stimuli entrain electromagnetic brain activity in such a way that an evoked

EEG/MEG response is produced with its frequency components remaining constant in amplitude and phase during the whole duration of sensory processing of presented stimuli. Unlike conventional ERPs/ERFs, which are analyzed primarily in the time domain (by calculating amplitude, latency, evaluating brain topography or estimating source localization at each particular time point), SSRs are evaluated in the frequency domain. The use of such responses may provide important insight into how internal brain activity and external information are synchronized in OCD. However, no ASSR studies in OCD have been conducted to date.

The situation is different for SCZ. A number of authors demonstrated deficits in ASSR generation in patients with SCZ as compared to HCS (for review, see 179]. The main findings related to ASSR abnormality in SCZ are: [1] a reduction of spectral power to 40-Hz clicks, in particular in the gamma-frequency band [180, 181]; [2] diminished inter-trial phase-locking [182]; and [3] delayed phase synchrony [181] when compared with HCS. However, some data points to the fact that under certain circumstances ASSRs may be augmented in SCZ [183] and might be resting state related [184].These findings imply an impaired mechanism of neuronal entrainment and possible alteration of synchronization/desynchronization mechanisms of electromagnetic brain activity in SCZ patients, especially in gamma-frequency band.

3.4. Early sensory processing

Sensory processing is altered in both OCD and SCZ patients. In OCD, it is can often be observed clinically as sensory intolerances or as a neurological soft sign [185]. A case series provide examples of pediatric patients with OCD who were significantly impaired by a chief complaint of a sensory intolerances to external environmental triggers, e.g. the sensation of oil on the skin, the smell of fish, salad dressing or cheese, and the tension of shoelaces and underwear [185-187]. In SCZ, on the other hand, the most frequent complaints of sensory changes are responses to internal visual or auditory experiences [188, 189]. Involvement of primary auditory areas in auditory hallucinations has been demonstrated in a number of studies with SCZ patients [190-192].

Early sensory information processing studied ERPs/ERFs in OCD. Sensory intolerances can be investigated on the neurophysiological level by studying early information processing with auditory, visual, or tactile stimuli. Indeed, a number of ERP studies demonstrated that processing of primary sensory information is altered in OCD patients. In our ERF study, Korostenskaja, Harris [193] demonstrated increased early auditory evoked fields M100 and M150 in the right hemisphere when compared to the right hemisphere in OCD subjects, whereas in no significant asymmetry was found in HCS. This interhemispheric asymmetry deserves detailed attention. This finding of increased auditory evoked response amplitudes over the right hemisphere is supported by other studies. In this way, Oades, Zerbin [194] found that OCD patients had higher N1 response amplitude in the right hemisphere, which was not the case in HCS. Morault et al. [195] showed that in response to verbal auditory stimuli presented in an "odd-ball" paradigm, patients with OCD had auditory evoked responses that are more positive in the left hemisphere while healthy subjects have more positive responses in the right hemisphere, however the opposite tendency was found for words when compared with HCS [196]. In addition, mismatch negativity (MMN) responses shifted to the right in subjects with OCD in a study by Oades et al. [194]. Gonçalves et al. [197] hypothesized that patients with OCD have an inter-hemispheric functional imbalance that is responsible for their symptoms and improves with treatment. Specifically, Serra et al. [85] suggested that patients with OCD have insufficient fronto-caudal regulation of the left hemisphere.

There is a large literature of support for interhemispheric asymmetry, supposedly reflecting interhemispheric dysfunction, in patients with OCD. Early evidence of such dysfunction in OCD subjects comes from neuropsychological performance results, implicating the dominant (left) hemisphere in the pathophysiology of OCD [86].

Anatomically, both gray matter and white matter interhemispheric differences have been found. There is reduced cortical folding in the left anterior cingulate cortex in subjects with OCD [198]. Using diffusion spectrum imaging, subjects with OCD were shown to have decreased left-lateralized asymmetry of the anterior segment of the cingulum bundles compared with HCS [199]. In pediatric patients with OCD, increased gray matter in the left putamen and right lateral orbitofrontal cortex correlate with OCD symptom severity [200].

Electrophysiologically, left hyperactivity in the frontal region is supported by alpha frequency bandwidth power increase [93]. Quantitative EEG analysis shows higher frequencies of slow-wave bands and lower frequencies of alpha activity predominantly in left frontotemporal regions in patients with OCD [82]. A left predominance of posterior frontal mid-temporal theta-2 was reported by [201].

Interhemispheric neurochemical differences have been found as well. Patients with OCD have higher binding ratios for the dopamine transporter in the left caudate and left putamen compared with healthy subjects and a higher D2 receptor binding potential in the left caudate [202, 203]. Interestingly, the P2 EEG response, which is considered to be an analogue of the M150 magnetic response has been proposed to reflect DA and NA activities [204]. M150 may reflect early stimulus evaluation and correspond to information inhibition processing in cortical and subcortical structures [205-207]. Importantly, deficits in inhibitory control were reported by a large number of studies on subjects with OCD [208, 209].

The data points to forward toward dissociation of early sensory processing deficits in auditory and visual modalities. Thus, early (before 200 ms) processing alterations in OCD patients are observed in the auditory, but not the visual, modality. For example, Savage, Weilburg [210] found shorter latencies for N1 and P2 responses of the auditory evoked potential to binaural clicks in adult OCD population. Importantly, the authors did not observe similar changes in the visual modality, suggesting particular involvement of auditory system in the pathophysiology of OCD. Moreover, the study by Ciesielski et al. [211] demonstrated that processing visual stimuli is altered in OCD subjects, only for the later but not the early stages of processing. In this way, the early visual component (P130) did not differ between HCS and OCD patients, but a later N220 component was reduced in amplitude and had shorter latency during the cognitive task consisting of presented two different picture stimuli in OCD subjects. However, one can expect changes already on the early stages of auditory processing in those OCD subjects, whose sensory intolerances are related to visual stimuli. More studies are needed to understand diverging results of processing auditory and visual information streams in OCD patients, including those with different types of sensory intolerances.

Early sensory information processing studied with ERPs/ERFs in SCZ. Although several studies reported early sensory auditory processing deficits in SCZ patients, which can start as early as after 15ms after the stimulus onset [212], the deficit at the early stages of visual information processing is more prominent and specific to this particular disorder.

Hypoactivation of the magnocellular pathway in patients with SCZ and schizoaffective disorder is a well documented phenomenon [213, 214]. Patients with SCZ have deficits at the early stages of visual information processing, starting with reduction of P1 amplitude to red light in VEPs [215] and following by reduced P1 during memory encoding and retrieval phases [216]. Notably, only a reduction in visual P1 amplitude but not in the later N1 and P2 components was found in SCZ patients in a study by Koychev et al. [217], although the study by Oribe et al. [218] demonstrated deficits on the late stages on visual information processing in SCZ patients as well.

Abnormalities in the latency and/or amplitude of auditory evoked responses may represent biomarkers of disease state in SCZ patients. Some investigators have shown that P2 component abnormalities represent physiological markers for a positive-symptom subtype [219, 220]. Likewise, Roth et al. [221] demonstrated a negative correlation between P2 latency to frequency tones and the delusion/hallucination score, and Laurent et al. [222] showed a negative correlation between P2 latency and the PANSS positive syndrome score, whereas Shenton et al. [223] found that reduced P2 amplitudes correlated significantly with a negative-symptom subtype.

In regards to functional asymmetry in SCZ, it found to be abnormal (for review, see 224, 225). Patients with SCZ failure to demonstrate functional asymmetry for language function. Although the main alterations in asymmetrical responses are observed at later stages of information processing, first changes can be detected as early as 150 ms [226]. Functional asymmetry in SCZ has been proposed to be utilized as a possible biomarker of SCZ disorder [227].

OCD and SCZ studies investigating early sensory information processing with ERPs/ERFs. No comparative studies investigating early stages of sensory processing in OCD and SCZ patients have been conducted to date. However, from the existing literature the following tendencies emerge: [1] Early auditory information processing is altered more OCD than in SCZ. This might be related to sensory intolerances that predominate in patients with OCD, more so than in patients with in SCZ; In OCD, such changes in auditory information processing can start already at the level of brain stem; [2] Early visual information processing is deficient in SCZ, but not in OCD; this tendency changes during the late stages of the processing, during which both OCD and SCZ show deficits. Comparative studies are needed for developing biomarkers distinguishing OCD and SCZ based on early visual and auditory ERP responses. Functional asymmetry can be a potential biomarker for these two psychiatric conditions.

3.5. Change detection processing

The auditory MMN, which can also be detected magnetically (MMN*m*), was first reported by Näätänen et al. [228] (see also, [229]). It is a negative ERP component elicited by any change in the repetitive auditory stimulus presentation (such as changes in frequency, duration, intensity, location) The MMN peaks at about 100-200 ms from change onset (for review, see [230]). It is suggested that the MMN represents a sensory memory trace formation process related to the evaluation of presented stimuli. The MMN could provide information about the amount of neuronal resources participating in automatic (involuntary) change-detection processes [231].

The change detection mechanism studied with MMN/MMNm in OCD. There are very few studies, to date, that investigate the change-detection mechanism in OCD patients. Most studies do not demonstrate significant differences between OCD patients relatively to HCS. For example, Towey et al. [232] did not find any alterations in MMN responses in OCD patients when compared with HCS. Nevertheless, they demonstrated differences between OCD patients and HCS in the later ERP components. It must be mentioned, though, that MMN in the above mentioned study was not elicited in the passive listening condition – without active participation, which is the most established approach to record this evoked response. Rather, the study participants were asked to pay active attention to deviant stimuli. Therefore, the study was addressing active rather than passive change-detection processes. More relevant information was obtained in a comparative MMN study between OCD and SCZ groups. This will be discussed below.

Perhaps it is unlikely for there to be abnormalities in the MMN in patients with OCD given its proposed neurochemical basis. The neurochemical dysfunction in OCD is thought by many to involve, deficits in serotonin (5-HT). Some studies confirm 5-HT involvement in MMN generation [233, 234]. However, it seems that NMDA-related changes have more influence on MMN generation than the serotonergic influences [235]. Therefore, one may not expect strong pronounced deficits in MMN generation in OCD patients.

Change detection mechanism studied with MMN/MMNm in SCZ. Unlike studies of MMN in OCD, the literature on MMN research in SCZ is very extensive. Both MMN and MMN*m* have proved to be particularly valuable in SCZ research (for review, see [236], [237]). The first report concerning MMN deficiency in SCZ was made by Shelley et al. [238]. Patients with SCZ show abnormal MMN and MMN*m* responses (for review, see [236], [239]) and the most significant finding is the reduction of the MMN amplitude [239, 240]. This is also shown in the magnetic MMN counterpart [241]. Important interhemispheric differences were also demonstrated in MMN response amplitude. Patients with SCZ seem to have lower MMN responses over the left hemisphere when compared with HCS [241, 242]. This corresponds well with MRI studies showing that SCZ patients have structural brain abnormalities with reduced grey matter density in the left posterior superior temporal gyrus, the medial temporal lobe structures [243], the left inferior parietal lobule, the cingulate gyrus, the left middle frontal gyrus, the left hippocampal gyrus and the right superior frontal cortex [244]. Moreover, the MMN amplitude in patients with SCZ correlates with the volume of primary auditory cortex (Heschl gyrus) [245]. In addition, several studies reported correlations between negative symptoms and the

MMN amplitude [240, 246]. More recently, studies of MMN in schizophrenia, utilize a functional connectivity analysis approach [247]. These authors demonstrated that cortical functional connectivity is impaired during "odd-ball" task eliciting MMN response. Imaginary coherence indexes measured from EEG activity in gamma frequency band between different brain regions correlated with hallucinatory behavior and clinical positive and negative symptom scores.

It is important to mention a relation between MMN generation and dysfunctional neurotransmission in SCZ. Glutamate NMDA receptors, which are thought to be implicated in pathophysiology of SCZ, are crucially involved in MMN generation [248, 249]. Another neurotransmitter system implicated in pathophysiology of SCZ was demonstrated to have significant inhibitory GABA influences on MMN generation as well [250, 251].

Comparative OCD and SCZ studies investigating change detection mechanism with MMN/MMNm. In their comparative study, Oades et al. [194] contrasted MMN findings between paranoid-hallucinatory and non-paranoid schizophrenia patients with OCD and HCS. Main differences were found in MMN scalp distribution. In this way, MMN amplitudes were higher on the right in OCD patients, whereas in a group of paranoid SCZ patients they were distributed bilaterally and in the group of non-paranoid SCZ patients they have shifted posteriorly. Right-hemispheric MMN amplitude asymmetry in OCD group is consistent with our previous findings of increased amplitudes of both M100 and M150 components over the right hemisphere in OCD patients [193]. In their study Oades et al. [252] found MMN reduced at frontal and increased at temporal sites of the brain in patients with SCZ in both passive and active attention conditions. Usually expected increase of MMN amplitude with switch to the active condition was observed only in OCD and HCS groups, but not in the SCZ group. Overall, both studies on neurochemical regulation of MMN and comparative studies between MMN responses in OCD and SCZ suggest the possibility of using the MMN as a marker to differentiate these disorders.

3.6. Attention

A specially designed "distraction" paradigm utilizes novel (distractive) stimuli to elicit P3a response [253, 254], which is considered to reflect reorienting of involuntary attention towards the novel (distracting) event. The P3a in EEG recordings is a positive response, peaking around 250-350 ms after stimulus onset [255]. It follows the processes reflected by it preceding MMN response [254]. The frontal lobes seem to be necessary for P3a generation, as P3a amplitude was significantly diminished in the presence of distracting stimuli in patients with frontal lobe lesions [256]. Similar to MMN, the P3a response can be recorded with MEG (P3a*m*) [257, 258]

The P300 (or P3) first described by Sutton et al. [259] is a positive potential occurring at an approximate latency of 300 ms and is evoked by the presentation of a deviant target (rare) stimulus embedded among irrelevant (frequent) stimuli, while the subject is actively reacting (pressing a button or mentally counting) to the target stimuli [260]. Classical P300 response requires positive response to the infrequent stimulus of an "odd-ball" task, has a parietal scalp maximum, and sometimes is referred to as P3b [261]. P300 is usually interpreted as an electrophysiological correlate of active attention processes and working memory [262]. The latency of P300 could correspond to the speed of cognitive processing or to that of stimulus classification [263]. It is notable that the P300 latency is negatively correlated with mental function in normal subjects, such that shorter latencies are related to superior cognitive performance [264].

Attention studied with P300 in OCD. There are few studies regarding novelty detection reflected in P3a response in patients with OCD. The only study we were able to find to date showed increased in novelty P3a amplitude in OCD patients compared with HCS [265]. It was interpreted by authors as "indicator of an enhanced cortical orienting response implicating stronger involuntary shifts of attention." Interestingly, the emotional context (neutral or negative) of stimulus presentation did not have any influence on P3a generation. It is worth mentioning here two other studies by Gohle, Juckel [266] and Mavrogiorgou, Juckel [267], who separated P3a and P3b subcomponents from P300 response elicited during classical "odd-ball" paradigm with one standard and one deviant stimuli in OCD patients. Study design did not utilize novel distracting stimuli here to elicit novelty P3a response.

Reports about alteration in P300 response in OCD patients are variable. Reduced P300 amplitude was demonstrated by Beech, Ciesielski [268] as well as by Towey, Tenke [232] in response to attended target stimuli. At the same time an increase in P300 amplitude was reported by other research groups [266, 267]. Interestingly, Towey, Tenke [232] observed differences in P300 amplitudes between two conditions – attended and unattended stimuli. P300 amplitude for unattended non-target stimuli was increased; at the same time it was decreased in response to attended targets. These authors speculated that these findings may imply abnormal allocation of attentional resources from relevant information (decreased P300 amplitude to the attended target stimulus) to irrelevant details (increased P300 amplitude to unattended non-target stimuli). Interestingly, the degree of P300 increase was shown to separate future treatment responders from non-responders [195]. Unlike the variable findings regarding P300 amplitude in OCD, the findings of changes in P300 latency are very consistent among the research groups. All studies show reduced P300 component latency in the OCD group when compared to HCS [195, 211, 267-270] (see also Table 2). This is very important finding demonstrating cortical hyperarousal associated with active attention processes and faster cognitive processes in OCD patients.

Attention studied with P300 in SCZ. Similar to MMN response, most of the published studies demonstrate decrease in P3a amplitude in SCZ patients when compared with HCS [271-273]. Reduction in P3a amplitude is strongly associated with clinical symptomatology, such as negative SCZ symptoms [272]. P3a reduction in SCZ is a well established phenomena and has also been confirmed in nonhuman primate model [274]. Together with MMN response, P3a changes are excellent proposals for biomarkers of SCZ [275, 276].

Patients with SCZ also show reduction of P300 amplitude, particularly in an auditory task [277-279]. Roth and Cannon [280] were the first to report reduced P300 amplitude in SCZ. Since that time, the reduction of P300 amplitude has been demonstrated in various experimental paradigms in acute, remitted, medicated and medication-free patients [277, 281-284]. Patients with SCZ also exhibit a delayed P300 latency [285, 286]. These effects are robust and independent of medication, gender, or clinical state at the time of testing. A positive correlation between the duration of schizophrenia illness and P300 latency was demonstrated [287]. A parietal P300 amplitude reduction in SCZhas been linkedto poorer performance on neuropsychological tests of memory, whereas frontal P300 amplitude reduction has been linked to impaired selective attention [288]. Notably, the relationship between neural P300 generator and clinical symptomatology was observed. In this way, Kim, Shim [289] demonstrated that the decreased P300 source activation in the middle temporal gyrus, posterior cingulate, precuneus, and superior occipital gyrus negatively correlated with negative symptom scores.

Comparative OCD and SCZ studies investigating attention processes with P300. Several comparative studies were conducted. Kim, Kang [290] compared P300 responses elicited by an auditory "odd-ball" paradigm in OCD, SCZ and a group of HCS; these authors also correlated neurophysiological results with neuropsychological scores. In this study, the P300 amplitude was smaller in both OCD and SCZ groups when compared with HCS. However, the differences in correlation between deficits in P300 generation and cognitive performance scores were observed in OCD and SCZ. Whereas P300-related cognitive deficits in OCD patients were localized and mostly related to controlled attention and self-guided behavior, the P300-associated cognitive deficits in SCZ were generalized, implying wide-range impairment. A more recent study by Pallanti, Castellini [291] explored not only differences in P300 responses between OCD and SCZ groups, but it also looked at SCZ patients exhibiting OCD behavior (Schizo-OCD patients). This group of patients demonstrated a distinct pattern of P300responses: Unlike OCD patients there was no differences in P300 responses between non-target and target conditions; unlike HCS there was elevated P300 amplitudes in the target condition and reduced P300 amplitudes in non-target condition. Thus it was possible to distinguish the SCZ-OCD patients from both the OCD and SCZ groups. These authors argue that using a neurophysiological approach one can separate a distinct clinical entity of Schizo-OCD from OCD and SCZ.

Overall, ERPs reflecting attention-influenced cognitive processes can be potential biomarkers distinguishing OCD and SCZ. Increased P3a amplitude in OCD and decreased P3a amplitude in SCZ, shorter P300 latencies in OCD and longer P300 latencies in SCZ may all be good candidates for making such a distinction. In addition, the, P300 can be used as a potential biomarker to distinguish Schizo-OCD subtype from both OCD and SCZ.

3.7. Action monitoring

Acton monitoring processes can be assessed by error related negativity (ERN, [292]) the negative portion of an event related potential that occurs 50-100 ms after a subject gives an incorrect response. The ERN is usually followed by a positive deflection or error positivity (PE, [293], [294]) and occurs 200-500 ms after an incorrect response.

Action monitoring studied with event-related negativity (ERN) in OCD. Patients with OCD are thought to monitor their actions excessively. Electrophysiological support for this notion comes from ERN measurements. Several studies have looked for error-related deviations in brain activity in subjects with OCD. Multiple studies have found greater ERN amplitude in subjects with OCD compared with healthy controls [292, 295-299]. Interestingly, Santesso et al., [300] expanded this finding to children. However, Nieuwenhuis et al. [301] did not find enhanced error-related activity in patients with OCD nor did Grundler et al. [296] in a population with subclinical obsessive compulsive symptoms using a probabilistic learning task. Yet, Grundler et al. [296] did find larger ERNs when using a flanker task. This finding of a differential task-dependent response was replicated by Endrass et al. [302].

Action monitoring studied with event-related negativity (ERN) in SCZ. In contradistinction to the results observed in OCD patients, the amplitude of ERN component related to action monitoring in SCZ patients is reduced [303-305]. This reduction was found to be associated with negative symptom severity and poorer real-world functioning (indicated by unemployment and re-hospitalization over 10 years of illness) in a study by Foti, Kotov [306]. Authors hypothesized that their results may represent decreased motivation to pursue goal-directed behavior, which is thought to underlie the exhibition of negative symptoms in SCZ.

Comparative studies on action monitoring with event-related negativity (ERN) between OCD and SCZ. To our knowledge, no comparative OCD-SCZ studies assessing action monitoring behavior with its neurophysiological analogue – ERN have been performed to date. From the individual reports of ERP studies in these two patient groups, it is evident that there are major differences in ERN generation. These differences includeincreased ERN amplitude in OCD and decreased ERN response in SCZ. These differences in ERN may be helpful as further biomarkers of disease.

In conclusion, there are obvious reasons to believe that neurophysiological markers distinguishing OCD and SCZ can be found. The main candidates are P3a and P3b responses as well as ERN. Additional studies are needed to determine whether there are changes in MMN in OCD. Future studies are recommended to dissociate deficits at the early stages of auditory and visual processes in OCD and SCZ. New studies evaluating entraining mechanisms with SSR in OCD are warranted. A special emphasis should be placed on examining developmental differences in the neurophysiological responses in OCD. The use of MEG is highly recommended in OCD group, as it can provide not only time-related information, but also localize the activity of interest in the space domain.

Studied processes	ERP/ERF component (-s) studied/paradigm	Study participants /age of OCD patients/	Observed ERP changes		Authors
		Brain stem function			
Brain stem function	BAEPs to clicks	50 OCD; 50 anxiety disorder; 25 HCS /adults; mean 33 ± 8 yo/	wave I–V wave III	interpeak L↑ A↓	Nolfe et al. [171]
		Early sensory processing (pre-attentive)			
Early visual processing	VEPs to flash	8 OCD; 8 HCS /adults; mean a ge 36.5 yo/	P130	-	Ciesielski et al. [211]
Early, auditory, visual and	AEPs to binaural clicks, VEPs to checkerboard flashes, SEPs	14 OCD; 14 neurotics; 14 HCS /adults/	Comparison with HCS:	A↑ A↓ A↓	Shagass et al. [324]

Studied processes	ERP/ERF component (-s) studied/paradigm	Study participants /age of OCD patients/	Observed ERP changes		Authors
somatosensory processing			Somatosensory P60 Visual N75 Auditory P50 Auditory P85	L↑	
Early somatosensory processing	SEPs	14 OCD; 28 neurotics; 99 HCS; 49 chronic SCZ; 27 "other" SCZ; 20 latent SCZ; 42 major depressive /adults; 23-48 yo/	Comparison with HCS: N60 P90 N130	A↑ A↓ A↓	Shagass et al. [325]
Early auditory and visual sensory processing	AEPs to tone bursts and VEPs to flash	50 OCD; 40 HCS /no age-related information was provided by authors/	Auditory and visual P1 N1 P2	-	Khanna et al. [326]
Early auditory and visual sensory processing	AEPs to binaural clicks and VEPs to flash	15 OCD (unmedicated); 30 HCS /adults; 35.6 ± 9.9/	Auditory N1 Auditory P2 Visual N1 Visual P2	L↓ L↓ - -	Savage et al. [210]
	BAEPs to	50 OCD; 50 anxiety disorder; 25 HCS /adults; mean 33 ± 8 yo/	wave I–V wave III		
Early auditory sensory processing	AEFs to binaural clicks	10 OCD; 10 HCS /youth; 8-13 yo/	M70 M100 M150	- L↑ -	Korostenskaja et al. [193]*
		Brain activity synchronization with external events (pre-attentive)			
Synchronation of brain activity with external periodic sensory stimulation	ASSR to repetitively presented trains of identical clicks	No studies reported to date in OCD patients; for comparison, see review section of ASSR studies in schizophrenia			
		Change detection (automatic attention processes)			
Automatic response to a change in external stimuli	Auditory MMN response to deviant tones interspersed among frequent tones. No task execution required	Only limited number of studies performed - see comparative OCD/SCZ section; for comparison, see also review section of MMN studies in schizophrenia			
		Novelty detection (involuntary attention switch)			
Processing of novel	AEPs elicited in response to auditory novelty "odd-ball" task with irrelevant	20 OCD; 20 HCS /adults; 32.8 ± 9.9 yo/	P3a	A↑	Ischebeck et al. [265]

Studied processes	ERP/ERF component (-s) studied/paradigm	Study participants /age of OCD patients/	Observed ERP changes		Authors
(unexpected) stimuli	repeated frequent sounds and rare novel sounds interspersed among them; the paradigm was presented during performance of visual recognition task				
		Active attention processes			
Selective attention; complex processing of visual information	VEPs elicited during visuospatial task: discrimination of two similar shapes	8 OCD; 8 HCS /adults; mean age 36.5 yo/	N220 P340	A↓; L↓ Similar trend as for N220, however not statistically significant	Ciesielski et al. [211]
Selective attention; complex processing of visual information	VEPs elicited during visuospatial task of increasing difficulty: discrimination of two similar shapes	8 OCD; 8 HCS /adults; average age 40 yo/	N220 P350	A ↓; L ↓ A ↓; L ↓ Effect was stronger with increasing task complexity	Beech et al. [268]
Active attention	AEPs elicited during auditory "odd-ball" discrimination paradigm of increasing difficulty	10 OCD (drug-free for two weeks); 10 HCS /adults; 18-55 yo/	N200 P300	A↑ L↓ during difficult discrimination task	Towey et al. [269]
Active attention	AEPs elicited during auditory "odd-ball" discrimination paradigm of increasing difficulty	17 OCD (unmedicated); 16 HCS /adults; 18-55 yo/	N200 P300	A↑ A over the left hemisphere "/> than over the right hemisphere -	Towey et al. [327]
Selective attention	AEPs elicited during direct attention task	18 OCD (unmedicated); 15 HCS /adults; mean age 30.0 ± 9.1 yo/	MMN (N2a) N200 (N2b) P300 for attended targets P300 for unattended non-targets	- - A↓ A↑ L↑, A↑ A↑	Towey et al. [232]

Studied processes	ERP/ERF component (-s) studied/paradigm	Study participants /age of OCD patients/	Observed ERP changes		Authors
			PN SW for unattended non-targets		
Active attention and discrimination of verbal stimuli	AEPs elicited to verbal (disyllabic) auditory stimuli (meaningful and meaningless) presented in an "oddball" paradigm; subjects were asked to keep a mental count of target stimuli	13 OCD (unmedicated); 13 HCS /adults; 21-56 yo/	N1 N2 P2 P3	L↑ A↓ L↑ L↓ A over the left hemisphere < than over the right hemisphere	Morault et al. [195]
Active attention and discrimination of auditory stimuli	AEPs elicited during auditory "odd-ball" paradigm with two frequency deviants; subjects were asked to keep a mental count of target stimuli	18 OCD (unmedicated); 18 HCS /adults; 19-59 yo/	N200 P300 SW	A↑ - L↓	de Groot et al. [329]
Active attention and discrimination of auditory stimuli	AEPs elicited during auditory "odd-ball" paradigm with two frequency deviants; subjects were asked to keep a mental count of target stimuli	23 OCD (unmedicated); 12 SP (unmedicated); 18 HCS 21 OCD (medication-free); 21 HCS / mixed: youth and adults; 16-50 yo/	N200 P300	L↓ than SP and HC A↑ than in HC L↓ than SP and HC	Miyata et al. [270]
Active attention	AEPs elicited during auditory "oddball" paradigm; authors do not specify details of the task performed by the subjects	30 OCD (medication-free for 2-4 weeks); 30 HCS /mixed: youth and adults; 16-40 yo/	P200 in response to irrelevant (frequent) stimuli N200	A↑ A↓	Okasha et al. [330]
Active attention	AEPs elicited during "oddball" paradigm with frequent and deviant (target) stimuli; subjects had to press a button in response to target stimuli	21 OCD (medication-free); 21 HCS / mixed: youth and adults; 17-27 yo/	P3a P3b (importantly, separation between P3a and P3b was performed with	- A↑ L ↓ (in the right hemisphere)	Mavrogiorgou et al. [267]

Studied processes	ERP/ERF component (-s) studied/paradigm	Study participants /age of OCD patients/	Observed ERP changes		Authors
			dipole modeling)		
Active attention	AEPs elicited during "oddball" paradigm with frequent and deviant (target) stimuli; subjects had to press a button in response to target stimuli	63 OCD (acutely ill, unmedicated); 63 HCS /adults; mean age 33.71 ± 10.17 yo/	P3a P3b (separation between P3a and P3b was performed with dipole modeling)	- A↑	Gohle et al. [266]
		Learning and memory			
Implicit memory; word repetition effect	Implicit memory (word repetition) task; ERPs to visually presented word/ non-word lexical decision task; subjects were asked to decide whether each item was word or non-word	12 OCD; 13 HCS /adults; 19-29 yo/	Word repetition effect at 300-500 ms post-stimulus Hemispheric functional asymmetry for the new words	OCD: No HCS: Yes OCD: Right-sided asymmetry HCS: Left-sided asymmetry	Kim et al. [331].
		Error-related behavior			
Action monitoring	ERPs were elicited during Stroop task, consisting of three visually presented words "red", "green", and "blue"; subjects were instructed to press the right or left mouse button in response to the color of the words	9 OCD; 9 HCS /adults; 19-58 yo/	ERN	A↑	Gehring et al. [292]
Action monitoring and target detection	ERP responses elicited during visual presentation of letters 'H' and 'O'; targets consisted of large letters, non-targets consisted of small letters; subjects were instructed to press a button held in the right hand whenever a	10 OCD; 10 HCS /adults; 22-40 yo/	ERN P3b	A↑, L↓ L↓	Johannes et al. [298]

Studied processes	ERP/ERF component (-s) studied/paradigm	Study participants /age of OCD patients/	Observed ERP changes		Authors
	target H appeared and to press another button in the left hand when a target O appeared				
Action monitoring and error detection	ERPs were elicited during Stroop task, consisting of three visually presented words "red", "green", and "blue"; subjects were instructed to press the right or left mouse button in response to the color of the words	18 high-OCD; 17 low-OCD	Response-locked ERN	A↑ in high-OCD group	[332]
Action monitoring	ERPs were recorded in probabilistic learning task and were associated with errors and negative feedback	16 OCD; 16 HCS /adults; 21-49 yo/	ERN	-	Nieuwenhuis et al. [301]
Error monitoring processes	ERPs were elicited during combined a Go/NoGo task with an Eriksen flanker paradigm	11 OCD; 11 HCS	ERN "early" Pe "late" Pe	A↑ - -	Ruchsow et al. [333]
Performance monitoring	ERPs were elicited during modified version of flanker interference task	20 OCD; 20 HCS /adults; 33.5 ± 8 yo/	ERN CRN Pe	A↑ A↑ -	Endrass et al. [295]
Performance monitoring before and after treatment	ERPs were elicited during modified version of Simon task	Before treatment: 18 OCD; 18 HCS After treatment: 10 OCD; 13 HCS /youth; 8 - 17 yo/	ERN	A↑ (the effect remained after treatment)	Hajcak et al. [297]
Performance monitoring	ERPs were elicited during modified version of flanker interference task	22 OCD; 22 HCS /adults; 31.2 ± 8.4 yo/	Standard condition: ERN CRN Pe	A↑ A↑ -	Endrass et al. [302]
Performance monitoring	ERPs was elicited during modified Erikson flankers task	25 OCD, 27 GAD, 27 HCS /adults; 32.5 ± 10.2 yo/	ERN Ne of difference waveform amplitude at	A↑ in OCD A↑ in OCD subjects, but not in GAD	Xiao et al. [334]

Studied processes	ERP/ERF component (-s) studied/paradigm	Study participants /age of OCD patients/	Observed ERP changes		Authors
			medial frontal electrodes Pe	(compared with HCS) -	
Performance monitoring	ERPs was elicited during modified Erikson flankers task	44 OCD; 44 HCS /youth; 10 - 19 yo/	ERN	A↑	Hanna et al. [335]
Performance monitoring	ERPs was elicited during modified Erikson flankers task	26 OCD; 13 non-OCD anxiety disorder; 27 HCS /youth; 8 - 16 yo/	ERN	A↑	Carrasco et al. [170]
Performance monitoring	ERPs was elicited during modified Erikson flankers task	40 OCD; 19 unaffected siblings of OCD subjects; 40 HCS /youth; 10 - 17 yo/	ERN	A↑ (in both OCD patients and their siblings)	Carrasco et al. [336]

A – amplitude; AEFs – auditory evoked fields; AEPs – auditory evoked potentials; ASSR – auditory steady-state evoked responses; BAEPs - Brainstem auditory evoked potentials; CNV – contingent negative variation; CRN – correct response negativity; ERN/Ne – error-related negativity; ERP – event related potential; GAD - generalized anxiety disorder; HAMA - Hamilton Anxiety Rating Scale; HAMD - Hamilton Depression Rating Scale; HCS – healthy comparison subjects; L – latency; MMN – mismatch negativity; OCD – obsessive-compulsive disorder; Pe – error positivity; PN – processing negativity; SCZ – schizophrenia; SEPs - sometosentory evoked potentials; SP - social phobia; SW – slow wave; T/C – test/conditioning ratio; VEPs – visual evoked potentials; yo – years old; ↓ – decrease; ↑ increase; "- " – no effect; * denotes MEG studies

Table 2. ERP/ERF studies in patients with OCD

4. Neurophysiological markers of treatment response

Neurophysiologically guided clinical decisions is an exciting direction for translational research. We are close to clinically useful predictors of treatment response in both OCD and SCZ. Finding such a biomarker or assessment battery has been a longstanding goal of clinicians and scientists, as the current treatment approach is one of trial-and-error with respect to choice of medication. Since both SCZ and OCD are chronic conditions, patients with life-long, treatment-resistant disorders suffer from complications including medication side effects, poor quality of life, depression, unemployment and stigma. Although the symptoms of OCD typically begin in childhood [307], only about half of youths with OCD respond to the current standard-of-care treatment consisting of a serotonergic medication and cognitive behavior therapy (POTS, [308]).

As early as 1984, Insel, Mueller [309] attempted to find a biomarker to predict treatment response. This group of investigators examined a number of tests, including sleep EEG, the dexamethasone suppression test and platelet 3H-imipramine binding, but they were unsuccessful in finding a marker that predicted clinical response to clomipramine in patients with OCD. More recently, a number of functional imaging studies in adults have described changes functional changes in specific anatomical regions as a response to treatment. Nakao et al. [317]

found that elevated activity in the OFC, dorsolateral PFC and anterior cingulate cortex decreased with fluvoxamine or cognitive behavior therapy. Rauch et al. [318] reported that lower pretreatment regional cerebral blood flow (rCBF) in the OFC and higher rCBF in the posterior cingulate cortex were predictive of treatment response to fluvoxamine in adults with contamination obsessions. Similarly, Saxena et al. [319] reported glucose metabolism decreased in the right anterolateral OFC and right caudate as a result of 8-12 weeks of treatment with paroxetine. Prior to treatment with clomipramine, Rubin et al. (1995) reported that there was increased uptake of HMPAO in the OFC, posterofrontal cortex and dorsal parietal cortex compared with healthy volunteers. Decreased uptake in these regions occurred following treatment.

With regard to electrophysiological studies, a series of three papers by a single team of investigators have identified two groups of patients with OCD; one group had elevated alpha power at baseline and the other group had elevated theta power. These groups predict treatment response: the majority of patients with excess alpha respond to paroxetine and the majority of the patients with excess theta were non-responders. The excess alpha observed in the resting state prior to treatment normalized with treatment [76-78]. Fontenelle et al. [95], on the other hand, found yet another EEG band predictive of treatment response. These investigators obtained the pretreatment EEG in 17 drug-free patients with OCD and analysed the EEG with low-resolution electromagnetic tomography. Subjects were then treated for 12 weeks with antidepressants; 10 subjects responded and 7 were considered non-responders. There was significantly lower beta band activity in the anterior cingulate and medial frontal gyrus in the pretreatment EEG of responders. In a randomized double-blind study comparing sham feedback with neurofeedback (NFB) for the treatment of OCD, a significant decrease in compulsions was seen in the NFB group. Unlike the findings of Fontanelle, these authors noted than an increase in delta, low alpha and low beta in the baseline EEG were predictive of worse treatment outcome [320].

In an early study of treatment prediction for SCZ, Galderisi et al., (1994) examined baseline QEEG characteristics and their changes following a single test dose of either haloperidol or clopenthixol in 29 patients with SCZ. Those who responded to medication less slow activity and more fast activity than nonresponders. However, the authors go on to note that there was overlap in the baseline activity of responders and non-responders decreasing the utility of this approach. Yet changes in alpha 1, observed 6 hours after the administration of a single test dose of either haloperidol or clopenthixol, did succeed in discriminating between responders and nonresponders. Antipsychotic medications currently used to treat SCZ symptoms In a more recent study of 22 drug-naïve patients with SCZ treated with a variety of medications including conventional dopamine-blocking neuroleptics, serotonin-dopamine antagonists, anticholinergic agents, antihistaminergic agents or benzodiazepine derivatives, no spectral changes were found when comparing EEGs pre and post-treatment. However, using a novel approach for treatment response assessment, this study used an analysis of multiscale entropy and found that subjects with SCZ had greater complexity for lower frequencies than HCS in fronto-centro-temporal regions, but not in parieto-occipital regions. Following treatment, the elevated complexity normalized in fronto-central regions but was not alleviated in temporal regions [315]. Based on these studies, one wonders whether pretreatment alpha power does not predict response to dopamine antagonists, but what about medications that target the glutamatergic system which can also alleviate psychotic symptoms in SCZ [316]? Clinical trials for such compounds may consider changes in alpha EEG activity as a biomarker of treatment response or an alternate approach to response prediction.

For patients with SCZ, Khodayari-Rostamabad et al. [310] successfully used an alternative analytic approach – that of machine learning-- to extract features in the pre-treatment EEG to predict clinical response to clozapine. As such novel analytic approaches gain popularity in neurophysiological research, there is hope that these will have translational value for treatment prediction in the future. One such illustrative example is the machine learning approach to develop computational models based on the patients' MMN*m* and clinical data [311]. In the field of epilepsy, investigators use machine learning to predict seizures outcome following neurosurgery with 90% success rate [312] and to predict the likelihood of having a seizure from EEG features [313]. New and important information about the effect of epilepsy on information processing was reported by Ralescu, Lee [311] (Figure 3). The advantage of the computational approach used by these authors is that it allows experimentation with various settings of the parameters to generate possible scenarios for different models [314]. This computational approach for psychophysiological data analysis may reveal individual patterns of activity within the group. This innovative solution may have a strong potential to provide new insights into predicting treatment response for other conditions using the neurophysiological parameters of EEG/MEG or ERP/ERF responses in both OCD and SCZ.

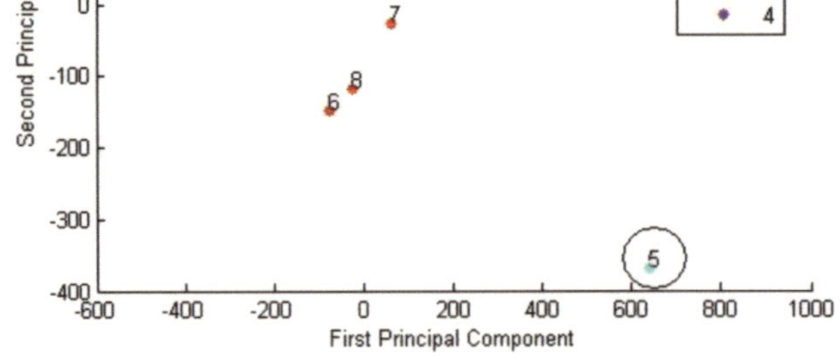

Figure 3. Clustering the MMN*m* response data of the ten patients with epilepsy in a two dimensional space of principal components. Patient P2 was farthest away from the rest of the data. Inspection of the P2 individual characteristics revealed that her age at onset of epilepsy (0.5 years) was the earliest among the rest of the patients. Utilized approach allowed differentiation of unique patients' characteristics through the parameters of neurophysiological MMN*m* responses.

5. Future directions

On the basis of clinical history and mental status examination, the young adult with unwinding, stealing and contamination would be given the diagnosis of OCD. The authors are hopeful that in the near future it will be possible to order an electrophysiological battery to confirm the diagnosis in challenging cases and to guide individually tailored treatment.

The hope that a biomarker for psychiatric conditions will emerge is already becoming a reality for some conditions. Basar's [321] proposal that brain functions are a result of simultaneous oscillations in various frequency bands has yielded fruit. Patients with ADHD can now be diagnosed based on the ratio of theta to beta frequency bandwidths. Robinson [322] found an inverse relationship between alpha and delta waves that correlated with personality type, with lower magnitude in extraverted and neurotic subjects. Changes in the cross-frequency coupling can be seen following treatment with psychotherapy [323]. Further examination of the interactions between different frequency bandwidths for patients with schizophrenia and OCD may be the logical next step.

Acknowledgements

The authors would like to acknowledge Mr. Brandon Lewis for his help in gathering a subset of the references for this review, Dr. Ernest Pedapati for his help with the TMS section, Dr. Mark DiFrancesco for Figure 1, Dr. Inga Griskova-Bulanova for providing useful comments on SSR section, and for support from the International OCD Foundation.

Author details

Elana Harris[1*] and Milena Korostenskaja[2,3]

*Address all correspondence to: elana.harris@cchmc.org

1 Division of Child and Adolescent Psychiatry, Cincinnati Children's Hospital Medical Center, Cincinnati, Ohio, USA

2 MEG Lab, Florida Hospital for Children, Orlando, FL, United States, USA

3 Functional Brain Mapping and Brain Computer Interface Lab, Florida Hospital for Children, Orlando, FL, USA

The authors have no financial or personal relationships to disclose.

References

[1] Weissman MM, Bland RC, Canino GJ, Greenwald S, Hwu HG, Lee CK, et al. The cross national epidemiology of obsessive compulsive disorder. The Cross National Collaborative Group. J Clin Psychiatry. 1994;55 Suppl:5-10. Epub 1994/03/01.

[2] Saha S, Chant D, McGrath J. Meta-analyses of the incidence and prevalence of schizophrenia: conceptual and methodological issues. International journal of methods in psychiatric research. 2008;17(1):55-61. Epub 2008/02/21.

[3] Saha S, Chant D, Welham J, McGrath J. A systematic review of the prevalence of schizophrenia. PLoS medicine. 2005;2(5):e141. Epub 2005/05/27.

[4] Sasson Y, Zohar J, Chopra M, Lustig M, Iancu I, Hendler T. Epidemiology of obsessive-compulsive disorder: a world view. J Clin Psychiatry. 1997;58 Suppl 12:7-10. Epub 1997/01/01.

[5] Whiteford HA, Degenhardt L, Rehm J, Baxter AJ, Ferrari AJ, Erskine HE, et al. Global burden of disease attributable to mental and substance use disorders: findings from the Global Burden of Disease Study 2010. Lancet. 2013. Epub 2013/09/03.

[6] Bobes J, Gonzalez MP, Bascaran MT, Arango C, Saiz PA, Bousono M. Quality of life and disability in patients with obsessive-compulsive disorder. Eur Psychiatry. 2001;16(4):239-45. Epub 2001/06/22.

[7] Huppert JD, Simpson HB, Nissenson KJ, Liebowitz MR, Foa EB. Quality of life and functional impairment in obsessive-compulsive disorder: a comparison of patients with and without comorbidity, patients in remission, and healthy controls. Depress Anxiety. 2009;26(1):39-45. Epub 2008/09/19.

[8] Bellack AS, Green MF, Cook JA, Fenton W, Harvey PD, Heaton RK, et al. Assessment of community functioning in people with schizophrenia and other severe mental illnesses: a white paper based on an NIMH-sponsored workshop. Schizophr Bull. 2007;33(3):805-22. Epub 2006/08/26.

[9] Caldwell CB, Gottesman, II. Schizophrenics kill themselves too: a review of risk factors for suicide. Schizophr Bull. 1990;16(4):571-89. Epub 1990/01/01.

[10] Eisen JL, Beer DA, Pato MT, Venditto TA, Rasmussen SA. Obsessive-compulsive disorder in patients with schizophrenia or schizoaffective disorder. Am J Psychiatry. 1997;154(2):271-3. Epub 1997/02/01.

[11] Berman I, Kalinowski A, Berman SM, Lengua J, Green AI. Obsessive and compulsive symptoms in chronic schizophrenia. Comprehensive psychiatry. 1995;36(1):6-10. Epub 1995/01/01.

[12] de Haan L, Dudek-Hodge C, Verhoeven Y, Denys D. Prevalence of psychotic disorders in patients with obsessive-compulsive disorder. CNS Spectr. 2009;14(8):415-7. Epub 2009/11/06.

[13] Anholt GE, Aderka IM, van Balkom AJ, Smit JH, Schruers K, van der Wee NJ, et al. Age of onset in obsessive-compulsive disorder: admixture analysis with a large sample. Psychol Med. 2013:1-10. Epub 2013/03/23.

[14] Hafner H, Maurer K, Loffler W, Fatkenheuer B, an der Heiden W, Riecher-Rossler A, et al. The epidemiology of early schizophrenia. Influence of age and gender on onset and early course. The British journal of psychiatry Supplement. 1994(23):29-38. Epub 1994/04/01.

[15] APA. Diagnostic and statistical manual of mental disorders, DSM-IV-tr (IV ed.). Washington, DC: American Psychiatric Pub Inc; 2000.

[16] Bora E, Murray RM. Meta-analysis of Cognitive Deficits in Ultra-high Risk to Psychosis and First-Episode Psychosis: Do the Cognitive Deficits Progress Over, or After, the Onset of Psychosis? Schizophr Bull. 2013. Epub 2013/06/19.

[17] Kerns JG, Berenbaum H. Cognitive impairments associated with formal thought disorder in people with schizophrenia. J Abnorm Psychol. 2002;111(2):211-24. Epub 2002/05/11.

[18] APA. Diagnostic and Statistical Manual of Mental Disorders, 5th Edition: DSM-5: American Psychiatric Publishing; 2013.

[19] Faragian S, Fuchs C, Pashinian A, Weizman R, Weizman A, Poyurovsky M. Age-of-onset of schizophrenic and obsessive-compulsive symptoms in patients with schizo-obsessive disorder. Psychiatry Res. 2012;197(1-2):19-22. Epub 2012/03/23.

[20] Bottas A, Cooke RG, Richter MA. Comorbidity and pathophysiology of obsessive-compulsive disorder in schizophrenia: is there evidence for a schizo-obsessive subtype of schizophrenia? J Psychiatry Neurosci. 2005;30(3):187-93. Epub 2005/06/10.

[21] Gross-Isseroff R, Hermesh H, Zohar J, Weizman A. Neuroimaging communality between schizophrenia and obsessive compulsive disorder: a putative basis for schizo-obsessive disorder? World J Biol Psychiatry. 2003;4(3):129-34. Epub 2003/07/23.

[22] Poyurovsky M, Weizman A, Weizman R. Obsessive-compulsive disorder in schizophrenia: clinical characteristics and treatment. CNS Drugs. 2004;18(14):989-1010. Epub 2004/12/09.

[23] Fuhr P, Agostino R, Hallett M. Spinal motor neuron excitability during the silent period after cortical stimulation. Electroencephalography and clinical neurophysiology. 1991;81(4):257-62. Epub 1991/08/01.

[24] Ziemann U, Netz J, Szelenyi A, Homberg V. Spinal and supraspinal mechanisms contribute to the silent period in the contracting soleus muscle after transcranial

magnetic stimulation of human motor cortex. Neuroscience letters. 1993;156(1-2): 167-71. Epub 1993/06/25.

[25] Kobayashi M, Pascual-Leone A. Transcranial magnetic stimulation in neurology. Lancet Neurol. 2003;2(3):145-56. Epub 2003/07/10.

[26] Bliss TV, Lomo T. Long-lasting potentiation of synaptic transmission in the dentate area of the anaesthetized rabbit following stimulation of the perforant path. J Physiol. 1973;232(2):331-56. Epub 1973/07/01.

[27] Pascual-Leone A, Amedi A, Fregni F, Merabet LB. The plastic human brain cortex. Annu Rev Neurosci. 2005;28:377-401. Epub 2005/07/19.

[28] Kujirai T, Caramia MD, Rothwell JC, Day BL, Thompson PD, Ferbert A, et al. Corticocortical inhibition in human motor cortex. J Physiol. 1993;471:501-19. Epub 1993/11/01.

[29] Ziemann U, Rothwell JC, Ridding MC. Interaction between intracortical inhibition and facilitation in human motor cortex. J Physiol. 1996;496 (Pt 3):873-81. Epub 1996/11/01.

[30] Di Lazzaro V, Restuccia D, Oliviero A, Profice P, Ferrara L, Insola A, et al. Magnetic transcranial stimulation at intensities below active motor threshold activates intracortical inhibitory circuits. Experimental brain research Experimentelle Hirnforschung. 1998;119(2):265-8. Epub 1998/04/16.

[31] Huang YZ, Edwards MJ, Rounis E, Bhatia KP, Rothwell JC. Theta burst stimulation of the human motor cortex. Neuron. 2005;45(2):201-6. Epub 2005/01/25.

[32] Bersani FS, Minichino A, Enticott PG, Mazzarini L, Khan N, Antonacci G, et al. Deep transcranial magnetic stimulation as a treatment for psychiatric disorders: a comprehensive review. Eur Psychiatry. 2013;28(1):30-9. Epub 2012/05/09.

[33] Kindler J, Homan P, Flury R, Strik W, Dierks T, Hubl D. Theta burst transcranial magnetic stimulation for the treatment of auditory verbal hallucinations: Results of a randomized controlled study. Psychiatry Res. 2013;209(1):114-7. Epub 2013/05/08.

[34] Kindler J, Homan P, Jann K, Federspiel A, Flury R, Hauf M, et al. Reduced neuronal activity in language-related regions after transcranial magnetic stimulation therapy for auditory verbal hallucinations. Biol Psychiatry. 2013;73(6):518-24. Epub 2012/07/31.

[35] Poulet E, Haesebaert F, Saoud M, Suaud-Chagny MF, Brunelin J. Treatment of shizophrenic patients and rTMS. Psychiatria Danubina. 2010;22 Suppl 1:S143-6. Epub 2011/02/08.

[36] Slotema CW, Blom JD, Hoek HW, Sommer IE. Should we expand the toolbox of psychiatric treatment methods to include Repetitive Transcranial Magnetic Stimulation

(rTMS)? A meta-analysis of the efficacy of rTMS in psychiatric disorders. J Clin Psychiatry. 2010;71(7):873-84. Epub 2010/04/07.

[37] Greenberg BD, Ziemann U, Harmon A, Murphy DL, Wassermann EM. Decreased neuronal inhibition in cerebral cortex in obsessive-compulsive disorder on transcranial magnetic stimulation. Lancet. 1998;352(9131):881-2. Epub 1998/09/22.

[38] Greenberg BD, Murphy DL, Rasmussen SA. Neuroanatomically based approaches to obsessive-compulsive disorder. Neurosurgery and transcranial magnetic stimulation. Psychiatr Clin North Am. 2000;23(3):671-86, xii. Epub 2000/09/15.

[39] Richter MA, de Jesus DR, Hoppenbrouwers S, Daigle M, Deluce J, Ravindran LN, et al. Evidence for cortical inhibitory and excitatory dysfunction in obsessive compulsive disorder. Neuropsychopharmacology. 2012;37(5):1144-51. Epub 2011/12/16.

[40] Pedapati EV, DiFranscesco M, Wu S, Giovanetti C, Nash T, Harris E. Functional Magnetic Imaging Reveals Neural Changes Associated With Transcranial Magnetic Stimulation In Adolescents With Obsessive Compulsive Disorder. American Academy of Child and Adolescent Psychaitry 60th Annual Meeting; October 22-27; Orlando, Florida, USA2013.

[41] Daskalakis ZJ, Christensen BK, Chen R, Fitzgerald PB, Zipursky RB, Kapur S. Evidence for impaired cortical inhibition in schizophrenia using transcranial magnetic stimulation. Arch Gen Psychiatry. 2002;59(4):347-54. Epub 2002/04/03.

[42] Fitzgerald PB, Brown TL, Daskalakis ZJ, Kulkarni J. A transcranial magnetic stimulation study of inhibitory deficits in the motor cortex in patients with schizophrenia. Psychiatry Res. 2002;114(1):11-22. Epub 2002/02/28.

[43] Fitzgerald PB, Brown TL, Marston NA, Oxley TJ, de Castella A, Daskalakis ZJ, et al. A transcranial magnetic stimulation study of abnormal cortical inhibition in schizophrenia. Psychiatry Res. 2003;118(3):197-207. Epub 2003/07/02.

[44] Eichhammer P, Wiegand R, Kharraz A, Langguth B, Binder H, Hajak G. Cortical excitability in neuroleptic-naive first-episode schizophrenic patients. Schizophr Res. 2004;67(2-3):253-9. Epub 2004/02/27.

[45] Liu SK, Fitzgerald PB, Daigle M, Chen R, Daskalakis ZJ. The relationship between cortical inhibition, antipsychotic treatment, and the symptoms of schizophrenia. Biol Psychiatry. 2009;65(6):503-9. Epub 2008/10/28.

[46] Wobrock T, Schneider-Axmann T, Retz W, Rosler M, Kadovic D, Falkai P, et al. Motor circuit abnormalities in first-episode schizophrenia assessed with transcranial magnetic stimulation. Pharmacopsychiatry. 2009;42(5):194-201. Epub 2009/09/03.

[47] Daskalakis ZJ, Christensen BK, Fitzgerald PB, Chen R. Dysfunctional neural plasticity in patients with schizophrenia. Arch Gen Psychiatry. 2008;65(4):378-85. Epub 2008/04/09.

[48] Wobrock T, Schneider M, Kadovic D, Schneider-Axmann T, Ecker UK, Retz W, et al. Reduced cortical inhibition in first-episode schizophrenia. Schizophr Res. 2008;105(1-3):252-61. Epub 2008/07/16.

[49] Fitzgerald PB, Brown TL, Daskalakis ZJ, Kulkarni J. A transcranial magnetic stimulation study of the effects of olanzapine and risperidone on motor cortical excitability in patients with schizophrenia. Psychopharmacology. 2002;162(1):74-81. Epub 2002/07/11.

[50] Daskalakis ZJ, Christensen BK, Chen R, Fitzgerald PB, Zipursky RB, Kapur S. Effect of antipsychotics on cortical inhibition using transcranial magnetic stimulation. Psychopharmacology. 2003;170(3):255-62. Epub 2003/08/09.

[51] Hamalainen MS, Hari R, Ilmoniemi RJ, Knuutila J, Lounasmaa OV. Magnetoencephalography - theory, instrumentation, and applications to noninvasive studies of theworking human brain. Rev Mod Phys. 1993;65:413-97.

[52] Cohen D. Magnetoencephalography: detection of the brain's electrical activity with a superconducting magnetometer. Science. 1972;175(4022):664-6. Epub 1972/02/11.

[53] Williams MA, Sachdev PS. Magnetoencephalography in neuropsychiatry: ready for application? Curr Opin Psychiatry. 2010;23(3):273-7. Epub 2010/03/11.

[54] Coburn KL, Lauterbach EC, Boutros NN, Black KJ, Arciniegas DB, Coffey CE. The value of quantitative electroencephalography in clinical psychiatry: a report by the Committee on Research of the American Neuropsychiatric Association. J Neuropsychiatry Clin Neurosci. 2006;18(4):460-500. Epub 2006/12/01.

[55] Hunter AM, Cook IA, Leuchter AF. The promise of the quantitative electroencephalogram as a predictor of antidepressant treatment outcomes in major depressive disorder. Psychiatr Clin North Am. 2007;30(1):105-24. Epub 2007/03/17.

[56] Takahashi T. Complexity of spontaneous brain activity in mental disorders. Progress in neuro-psychopharmacology & biological psychiatry. 2013;45:258-66. Epub 2012/05/15.

[57] Schmitt A, Hasan A, Gruber O, Falkai P. Schizophrenia as a disorder of disconnectivity. Eur Arch Psychiatry Clin Neurosci. 2011;261 Suppl 2:S150-4. Epub 2011/08/26.

[58] Hinkley LB, Vinogradov S, Guggisberg AG, Fisher M, Findlay AM, Nagarajan SS. Clinical symptoms and alpha band resting-state functional connectivity imaging in patients with schizophrenia: implications for novel approaches to treatment. Biol Psychiatry. 2011;70(12):1134-42. Epub 2011/08/25.

[59] Vrba J, Robinson SE. Signal processing in magnetoencephalography. Methods. 2001;25(2):249-71. Epub 2002/01/29.

[60] Baillet S, Friston K, Oostenveld R. Academic software applications for electromagnetic brain mapping using MEG and EEG. Computational intelligence and neuroscience. 2011;2011:972050. Epub 2011/08/09.

[61] Cohen D, Cuffin BN. EEG versus MEG localization accuracy: theory and experiment. Brain Topogr. 1991;4(2):95-103. Epub 1991/01/01.

[62] Cuffin BN. Effects of head shape on EEG's and MEG's. IEEE Trans Biomed Eng. 1990;37(1):44-52. Epub 1990/01/01.

[63] Ahlfors SP, Han J, Belliveau JW, Hamalainen MS. Sensitivity of MEG and EEG to source orientation. Brain Topogr. 2010;23(3):227-32. Epub 2010/07/20.

[64] Melcher JR, Cohen D. Dependence of the MEG on dipole orientation in the rabbit head. Electroencephalography and clinical neurophysiology. 1988;70(5):460-72. Epub 1988/11/01.

[65] Haueisen J, Ramon C, Czapski P, Eiselt M. On the influence of volume currents and extended sources on neuromagnetic fields: a simulation study. Annals of biomedical engineering. 1995;23(6):728-39. Epub 1995/11/01.

[66] Silverman JS, Loychik SG. Brain-mapping abnormalities in a family with three obsessive compulsive children. J Neuropsychiatry Clin Neurosci. 1990;2(3):319-22. Epub 1990/01/01.

[67] Alonso P, Orbegozo A, Pujol J, Lopez-Sola C, Fullana MA, Segalas C, et al. Neural correlates of obsessive-compulsive related dysfunctional beliefs. Progress in neuropsychopharmacology & biological psychiatry. 2013. Epub 2013/08/06.

[68] Cottraux J, Gerard D, Cinotti L, Froment JC, Deiber MP, Le Bars D, et al. A controlled positron emission tomography study of obsessive and neutral auditory stimulation in obsessive-compulsive disorder with checking rituals. Psychiatry Res. 1996;60(2-3): 101-12. Epub 1996/03/29.

[69] Hillebrand A, Barnes GR. A quantitative assessment of the sensitivity of whole-head MEG to activity in the adult human cortex. NeuroImage. 2002;16(3 Pt 1):638-50. Epub 2002/08/10.

[70] Northoff G, Richter A, Gessner M, Schlagenhauf F, Fell J, Baumgart F, et al. Functional dissociation between medial and lateral prefrontal cortical spatiotemporal activation in negative and positive emotions: a combined fMRI/MEG study. Cereb Cortex. 2000;10(1):93-107. Epub 2000/01/20.

[71] Amo C, Quesney LF, Ortiz T, Maestu F, Fernandez A, Lopez-Ibor MI, et al. Limbic paroxysmal magnetoencephalographic activity in 12 obsessive-compulsive disorder patients: a new diagnostic finding. J Clin Psychiatry. 2004;65(2):156-62. Epub 2004/03/09.

[72] Kringelbach ML, Lehtonen A, Squire S, Harvey AG, Craske MG, Holliday IE, et al. A specific and rapid neural signature for parental instinct. PLoS ONE. 2008;3(2):e1664. Epub 2008/02/28.

[73] Hughes JR, John ER. Conventional and quantitative electroencephalography in psychiatry. J Neuropsychiatry Clin Neurosci. 1999;11(2):190-208. Epub 1999/05/20.

[74] Pacella B, Polation P, Nagle S. Clinical and EEG studies in obsessive compulsive states. American Journal of Psychiatry. 1944;100:830-8.

[75] Insel TR, Donnelly EF, Lalakea ML, Alterman IS, Murphy DL. Neurological and neuropsychological studies of patients with obsessive-compulsive disorder. Biol Psychiatry. 1983;18(7):741-51. Epub 1983/07/01.

[76] Bolwig TG, Hansen ES, Hansen A, Merkin H, Prichep LS. Toward a better understanding of the pathophysiology of OCD SSRI responders: QEEG source localization. Acta Psychiatr Scand. 2007;115(3):237-42. Epub 2007/02/17.

[77] Hansen ES, Prichep LS, Bolwig TG, John ER. Quantitative electroencephalography in OCD patients treated with paroxetine. Clin Electroencephalogr. 2003;34(2):70-4. Epub 2003/06/06.

[78] Prichep LS, Mas F, Hollander E, Liebowitz M, John ER, Almas M, et al. Quantitative electroencephalographic subtyping of obsessive-compulsive disorder. Psychiatry Res. 1993;50(1):25-32. Epub 1993/04/01.

[79] Velikova S, Locatelli M, Insacco C, Smeraldi E, Comi G, Leocani L. Dysfunctional brain circuitry in obsessive-compulsive disorder: source and coherence analysis of EEG rhythms. NeuroImage. 2010;49(1):977-83. Epub 2009/08/18.

[80] Bucci P, Mucci A, Volpe U, Merlotti E, Galderisi S, Maj M. Executive hypercontrol in obsessive-compulsive disorder: electrophysiological and neuropsychological indices. Clin Neurophysiol. 2004;115(6):1340-8. Epub 2004/05/12.

[81] Pogarell O, Juckel G, Mavrogiorgou P, Mulert C, Folkerts M, Hauke W, et al. Symptom-specific EEG power correlations in patients with obsessive-compulsive disorder. Int J Psychophysiol. 2006;62(1):87-92. Epub 2006/03/24.

[82] Tot S, Ozge A, Comelekoglu U, Yazici K, Bal N. Association of QEEG findings with clinical characteristics of OCD: evidence of left frontotemporal dysfunction. Can J Psychiatry. 2002;47(6):538-45. Epub 2002/09/06.

[83] Locatelli M, Bellodi L, Grassi B, Scarone S. EEG power modifications in obsessive-compulsive disorder during olfactory stimulation. Biol Psychiatry. 1996;39(5):326-31. Epub 1996/03/01.

[84] Karadag F, Oguzhanoglu NK, Kurt T, Oguzhanoglu A, Atesci F, Ozdel O. Quantitative EEG analysis in obsessive compulsive disorder. The International journal of neuroscience. 2003;113(6):833-47. Epub 2003/05/31.

[85] Serra FP, Palma V, Nolfe G, Buscaino GA. An electrophysiological study in obsessional compulsive disorders. Acta neurologica. 1994;16(5-6):240-8. Epub 1994/12/01.

[86] Flor-Henry P, Yeudall LT, Koles ZJ, Howarth BG. Neuropsychological and power spectral EEG investigations of the obsessive-compulsive syndrome. Biol Psychiatry. 1979;14(1):119-30. Epub 1979/02/01.

[87] Kuskowski MA, Malone SM, Kim SW, Dysken MW, Okaya AJ, Christensen KJ. Quantitative EEG in obsessive-compulsive disorder. Biol Psychiatry. 1993;33(6): 423-30. Epub 1993/03/15.

[88] Khanna S. Obsessive-compulsive disorder: is there a frontal lobe dysfunction? Biol Psychiatry. 1988;24(5):602-13. Epub 1988/09/01.

[89] Jenike MA, Brotman AW. The EEG in obsessive-compulsive disorder. J Clin Psychiatry. 1984;45(3):122-4. Epub 1984/03/01.

[90] Olbrich S, Olbrich H, Adamaszek M, Jahn I, Hegerl U, Stengler K. Altered EEG lagged coherence during rest in obsessive-compulsive disorder. Clin Neurophysiol. 2013. Epub 2013/08/24.

[91] Koprivova J, Horacek J, Raszka M, Brunovsky M, Prasko J. Standardized low-resolution electromagnetic tomography in obsessive-compulsive disorder--a replication study. Neuroscience letters. 2013;548:185-9. Epub 2013/05/25.

[92] Desarkar P, Sinha VK, Jagadheesan K, Nizamie SH. Subcortical functioning in obsessive-compulsive disorder: an exploratory EEG coherence study. World J Biol Psychiatry. 2007;8(3):196-200. Epub 2007/07/27.

[93] Shin YW, Ha TH, Kim SY, Kwon JS. Association between EEG alpha power and visuospatial function in obsessive-compulsive disorder. Psychiatry Clin Neurosci. 2004;58(1):16-20. Epub 2003/12/18.

[94] Leocani L, Toro C, Zhuang P, Gerloff C, Hallett M. Event-related desynchronization in reaction time paradigms: a comparison with event-related potentials and corticospinal excitability. Clin Neurophysiol. 2001;112(5):923-30. Epub 2001/05/05.

[95] Fontenelle L, Piedade RA, Marques C, de Menezes GB, Versiani M. [Quantitative electroencephalography in obsessive-compulsive disorder: preliminary results]. Arquivos de neuro-psiquiatria. 2000;58(3A):677-82. Epub 2000/09/06. Eletrencefalografia quantitativa no transtorno obsessivo-compulsivo: resultados preliminares.

[96] Simpson HB, Tenke CE, Towey JB, Liebowitz MR, Bruder GE. Symptom provocation alters behavioral ratings and brain electrical activity in obsessive-compulsive disorder: a preliminary study. Psychiatry Res. 2000;95(2):149-55. Epub 2000/08/30.

[97] Amo C, Fernandez A, Leon JM, Ortiz T, Maestu F, Ferre F, et al. Paroxysmal MEG activity in obsessive compulsive patients without SSRIs therapy. Eur Psychiatry. 2006;21(2):139-41. Epub 2006/03/07.

[98] Maihofner C, Sperling W, Kaltenhauser M, Bleich S, de Zwaan M, Wiltfang J, et al. Spontaneous magnetoencephalographic activity in patients with obsessive-compulsive disorder. Brain Res. 2007;1129(1):200-5. Epub 2006/12/13.

[99] Ciesielski KT, Hamalainen MS, Lesnik PG, Geller DA, Ahlfors SP. Increased MEG activation in OCD reflects a compensatory mechanism specific to the phase of a visual working memory task. NeuroImage. 2005;24(4):1180-91. Epub 2005/01/27.

[100] Ciesielski KT, Hamalainen MS, Geller DA, Wilhelm S, Goldsmith TE, Ahlfors SP. Dissociation between MEG alpha modulation and performance accuracy on visual working memory task in obsessive compulsive disorder. Human brain mapping. 2007;28(12):1401-14. Epub 2007/03/21.

[101] Harris E, Robinson S, Holroyd T, Xiang J, Rose D, Horn P, et al. Timing of prefrontal cortical activation in the beta frequency bandwidth in adolescents with obsessive compulsive disorder. The 19th Annual International OCD (Obsessive-Compulsive Disorder) Foundation Conference (ICODF) July 27–29, 2012; Chicago, IL, USA2012.

[102] Harris E, Robinson S, Holroyd T, Rose D, Xiang J, Korostenskaja M, et al. A magnetoencephalographic study of symptom provocation in youth with obsessive compulsive disorder. Society for Neuroscience, 2012; October 13-17, 2012; New Orleans, LA, USA2012. p. 168.07.

[103] Gilbert AR, Akkal D, Almeida JR, Mataix-Cols D, Kalas C, Devlin B, et al. Neural correlates of symptom dimensions in pediatric obsessive-compulsive disorder: a functional magnetic resonance imaging study. J Am Acad Child Adolesc Psychiatry. 2009;48(9):936-44. Epub 2009/07/25.

[104] Hirosawa R, Narumoto J, Sakai Y, Nishida S, Ishida T, Nakamae T, et al. Reduced dorsolateral prefrontal cortical hemodynamic response in adult obsessive-compulsive disorder as measured by near-infrared spectroscopy during the verbal fluency task. Neuropsychiatric disease and treatment. 2013;9:955-62. Epub 2013/07/23.

[105] Menzies L, Chamberlain SR, Laird AR, Thelen SM, Sahakian BJ, Bullmore ET. Integrating evidence from neuroimaging and neuropsychological studies of obsessive-compulsive disorder: the orbitofronto-striatal model revisited. Neurosci Biobehav Rev. 2008;32(3):525-49. Epub 2007/12/07.

[106] Lemere F. The significance of individual differences in the Berger rhythm. Brain. 1936;59:366-75.

[107] Berger H. On the electroencephalogram of man. Twelfth report. Archiv fur Psychiatrie und Nervenkrankheiten. 1937;106:165-87.

[108] Ellingson RJ. The incidence of EEG abnormality among patients with mental disorders of apparently nonorganic origin: a critical review. Am J Psychiatry. 1954;111(4): 263-75. Epub 1954/10/01.

[109] Hill D. Clinical applications of EEG in psychiatry. The Journal of mental science. 1956;102(427):264-71. Epub 1956/04/01.

[110] Itil TM, Saletu B, Davis S. EEG findings in chronic schizophrenics based on digital computer period analysis and analog power spectra. Biol Psychiatry. 1972;5(1):1-13. Epub 1972/08/01.

[111] Gibbs FA, Gibbs EL, Lennox WG. The likeness of the cortical dysrhythmias of schizophrenia and psychomotor epilepsy. American Journal of Psychiatry. 1938;95:255-69.

[112] Davis PA, Davis H. The electroencephalograms of psychotic patients. American Journal of Psychiatry. 1939;95:1007-25.

[113] Jasper HH, Fitzpatrick CP, Solomon P. Analogies and opposites in schizophrenia and epilepsy. American Journal of Psychiatry. 1939;95:835-51.

[114] Finley KH, Campbell CM. Electroencephalography in schizophrenia. American Journal of Psychiatry. 1941;98:374-81.

[115] Greenblatt M. Age and electroencephalographic abnormality in neuropsychiatric patients: A study of 1,593 cases. American Journal of Psychiatry. 1944;101:82-90.

[116] Ostow M, Ostow M. Bilaterally synchronous paroxysmal slow activity in the encephalograms of non-epileptics. Journal of Nervous and Mental Disease. 1946;103:346-58.

[117] Bonkalo A, Doust JW, Stokes AB. Physiological concomitants of the phasic disturbances seen in periodic catatonia. Am J Psychiatry. 1955;112(2):114-22. Epub 1955/08/01.

[118] MacMahon JF, Walter WG. The electroencephalogram in schizophrenia. Journal of Mental Science. 1938;84:781-7.

[119] Colony HS, Willis SE. Electroencephalographic studies of 1,000 schizophrenic patients. Am J Psychiatry. 1956;113(2):163-9. Epub 1956/08/01.

[120] Hurst LA. Electroencephalographic support for a genetically oriented organic concept of schizophrenia. The Journal of nervous and mental disease. 1952;115(2):95-120. Epub 1952/02/01.

[121] Kennard MA, Levy S. The meaning of the abnormal electro-encephalogram in schizophrenia. The Journal of nervous and mental disease. 1952;116(5):413-25. Epub 1952/11/01.

[122] Kammerer T, Rohmer F, Wackerheim A. L'EEG des Schizophrenes. Strasbourg Medical. 1955;10 (Suppl.):20.

[123] Sponheim SR, Clementz BA, Iacono WG, Beiser M. Clinical and biological concomitants of resting state EEG power abnormalities in schizophrenia. Biol Psychiatry. 2000;48(11):1088-97. Epub 2000/11/30.

[124] Sponheim SR, Iacono WG, Clementz BA, Beiser M. Season of birth and electroencephalogram power abnormalities in schizophrenia. Biol Psychiatry. 1997;41(10): 1020-7. Epub 1997/05/15.

[125] Shagass C, Roemer RA, Straumanis JJ, Josiassen RC. Psychiatric diagnostic discriminations with combinations of quantitative EEG variables. Br J Psychiatry. 1984;144:581-92. Epub 1984/06/01.

[126] Gerez M, Tello A. Selected quantitative EEG (QEEG) and event-related potential (ERP) variables as discriminators for positive and negative schizophrenia. Biol Psychiatry. 1995;38(1):34-49. Epub 1995/07/01.

[127] Sponheim SR, Iacono WG, Thuras PD, Nugent SM, Beiser M. Sensitivity and specificity of select biological indices in characterizing psychotic patients and their relatives. Schizophr Res. 2003;63(1-2):27-38. Epub 2003/08/02.

[128] Clementz BA, Sponheim SR, Iacono WG, Beiser M. Resting EEG in first-episode schizophrenia patients, bipolar psychosis patients, and their first-degree relatives. Psychophysiology. 1994;31(5):486-94. Epub 1994/09/01.

[129] Venables NC, Bernat EM, Sponheim SR. Genetic and disorder-specific aspects of resting state EEG abnormalities in schizophrenia. Schizophr Bull. 2009;35(4):826-39. Epub 2008/04/03.

[130] Karson CN, Coppola R, Daniel DG, Weinberger DR. Computerized EEG in schizophrenia. Schizophr Bull. 1988;14(2):193-7. Epub 1988/01/01.

[131] Omori M, Koshino Y, Murata T, Murata I, Nishio M, Sakamoto K, et al. Quantitative EEG in never-treated schizophrenic patients. Biol Psychiatry. 1995;38(5):305-9. Epub 1995/09/01.

[132] Merrin EL, Floyd TC. Negative symptoms and EEG alpha in schizophrenia: a replication. Schizophr Res. 1996;19(2-3):151-61. Epub 1996/05/01.

[133] Hamm JP, Gilmore CS, Picchetti NA, Sponheim SR, Clementz BA. Abnormalities of neuronal oscillations and temporal integration to low- and high-frequency auditory stimulation in schizophrenia. Biol Psychiatry. 2011;69(10):989-96. Epub 2011/01/11.

[134] Clementz BA, Keil A, Kissler J. Aberrant brain dynamics in schizophrenia: delayed buildup and prolonged decay of the visual steady-state response. Brain Res Cogn Brain Res. 2004;18(2):121-9. Epub 2004/01/23.

[135] Hong LE, Summerfelt A, Mitchell BD, O'Donnell P, Thaker GK. A shared low-frequency oscillatory rhythm abnormality in resting and sensory gating in schizophrenia. Clin Neurophysiol. 2012;123(2):285-92. Epub 2011/08/25.

[136] Ferrarelli F, Tononi G. The thalamic reticular nucleus and schizophrenia. Schizophr Bull. 2011;37(2):306-15. Epub 2010/12/07.

[137] Llinas RR, Ribary U, Jeanmonod D, Kronberg E, Mitra PP. Thalamocortical dysrhythmia: A neurological and neuropsychiatric syndrome characterized by magnetoencephalography. Proc Natl Acad Sci U S A. 1999;96(26):15222-7. Epub 1999/12/28.

[138] Klimesch W, Sauseng P, Hanslmayr S. EEG alpha oscillations: the inhibition-timing hypothesis. Brain Res Rev. 2007;53(1):63-88. Epub 2006/08/05.

[139] Siekmeier PJ, Stufflebeam SM. Patterns of spontaneous magnetoencephalographic activity in patients with schizophrenia. J Clin Neurophysiol. 2010;27(3):179-90. Epub 2010/05/13.

[140] Boutros NN, Arfken C, Galderisi S, Warrick J, Pratt G, Iacono W. The status of spectral EEG abnormality as a diagnostic test for schizophrenia. Schizophr Res. 2008;99(1-3):225-37. Epub 2007/12/28.

[141] Dias EC, Bickel S, Epstein ML, Sehatpour P, Javitt DC. Abnormal task modulation of oscillatory neural activity in schizophrenia. Frontiers in psychology. 2013;4:540. Epub 2013/08/30.

[142] Uhlhaas PJ, Singer W. Abnormal neural oscillations and synchrony in schizophrenia. Nature reviews Neuroscience. 2010;11(2):100-13. Epub 2010/01/21.

[143] Lewis DA, Hashimoto T, Volk DW. Cortical inhibitory neurons and schizophrenia. Nature reviews Neuroscience. 2005;6(4):312-24. Epub 2005/04/02.

[144] Spencer KM, Nestor PG, Perlmutter R, Niznikiewicz MA, Klump MC, Frumin M, et al. Neural synchrony indexes disordered perception and cognition in schizophrenia. Proc Natl Acad Sci U S A. 2004;101(49):17288-93. Epub 2004/11/18.

[145] Gallinat J, Winterer G, Herrmann CS, Senkowski D. Reduced oscillatory gamma-band responses in unmedicated schizophrenic patients indicate impaired frontal network processing. Clin Neurophysiol. 2004;115(8):1863-74. Epub 2004/07/21.

[146] Symond MP, Harris AW, Gordon E, Williams LM. "Gamma synchrony" in first-episode schizophrenia: a disorder of temporal connectivity? Am J Psychiatry. 2005;162(3):459-65. Epub 2005/03/03.

[147] Wynn JK, Light GA, Breitmeyer B, Nuechterlein KH, Green MF. Event-related gamma activity in schizophrenia patients during a visual backward-masking task. Am J Psychiatry. 2005;162(12):2330-6. Epub 2005/12/07.

[148] Ford JM, Roach BJ, Faustman WO, Mathalon DH. Out-of-synch and out-of-sorts: dysfunction of motor-sensory communication in schizophrenia. Biol Psychiatry. 2008;63(8):736-43. Epub 2007/11/06.

[149] Cho RY, Konecky RO, Carter CS. Impairments in frontal cortical gamma synchrony and cognitive control in schizophrenia. Proc Natl Acad Sci U S A. 2006;103(52): 19878-83. Epub 2006/12/16.

[150] Lee SH, Wynn JK, Green MF, Kim H, Lee KJ, Nam M, et al. Quantitative EEG and low resolution electromagnetic tomography (LORETA) imaging of patients with persistent auditory hallucinations. Schizophr Res. 2006;83(2-3):111-9. Epub 2006/03/10.

[151] Spencer KM, Niznikiewicz MA, Nestor PG, Shenton ME, McCarley RW. Left auditory cortex gamma synchronization and auditory hallucination symptoms in schizophrenia. BMC Neurosci. 2009;10:85. Epub 2009/07/22.

[152] Mulert C, Kirsch V, Pascual-Marqui R, McCarley RW, Spencer KM. Long-range synchrony of gamma oscillations and auditory hallucination symptoms in schizophrenia. Int J Psychophysiol. 2011;79(1):55-63. Epub 2010/08/18.

[153] Bassett DS, Bullmore ET. Human brain networks in health and disease. Curr Opin Neurol. 2009;22(4):340-7. Epub 2009/06/06.

[154] Siebenhuhner F, Weiss SA, Coppola R, Weinberger DR, Bassett DS. Intra- and interfrequency brain network structure in health and schizophrenia. PLoS ONE. 2013;8(8):e72351. Epub 2013/08/31.

[155] Stephan KE, Friston KJ, Frith CD. Dysconnection in schizophrenia: from abnormal synaptic plasticity to failures of self-monitoring. Schizophr Bull. 2009;35(3):509-27. Epub 2009/01/22.

[156] Wang XJ. Neurophysiological and computational principles of cortical rhythms in cognition. Physiological reviews. 2010;90(3):1195-268. Epub 2010/07/29.

[157] Bassett DS, Bullmore E, Verchinski BA, Mattay VS, Weinberger DR, Meyer-Lindenberg A. Hierarchical organization of human cortical networks in health and schizophrenia. J Neurosci. 2008;28(37):9239-48. Epub 2008/09/12.

[158] Bassett DS, Bullmore ET, Meyer-Lindenberg A, Apud JA, Weinberger DR, Coppola R. Cognitive fitness of cost-efficient brain functional networks. Proc Natl Acad Sci U S A. 2009;106(28):11747-52. Epub 2009/07/01.

[159] Bassett DS, Nelson BG, Mueller BA, Camchong J, Lim KO. Altered resting state complexity in schizophrenia. NeuroImage. 2012;59(3):2196-207. Epub 2011/10/20.

[160] Liu Y, Liang M, Zhou Y, He Y, Hao Y, Song M, et al. Disrupted small-world networks in schizophrenia. Brain. 2008;131(Pt 4):945-61. Epub 2008/02/27.

[161] Lynall ME, Bassett DS, Kerwin R, McKenna PJ, Kitzbichler M, Muller U, et al. Functional connectivity and brain networks in schizophrenia. J Neurosci. 2010;30(28): 9477-87. Epub 2010/07/16.

[162] Weiss SA, Bassett DS, Rubinstein D, Holroyd T, Apud J, Dickinson D, et al. Functional Brain Network Characterization and Adaptivity during Task Practice in Healthy Volunteers and People with Schizophrenia. Front Hum Neurosci. 2011;5:81. Epub 2011/09/03.

[163] Rubinov M, Sporns O. Complex network measures of brain connectivity: uses and interpretations. NeuroImage. 2010;52(3):1059-69. Epub 2009/10/13.

[164] Fornito A, Zalesky A, Pantelis C, Bullmore ET. Schizophrenia, neuroimaging and connectomics. NeuroImage. 2012;62(4):2296-314. Epub 2012/03/06.

[165] Zalesky A, Fornito A, Egan GF, Pantelis C, Bullmore ET. The relationship between regional and inter-regional functional connectivity deficits in schizophrenia. Human brain mapping. 2012;33(11):2535-49. Epub 2011/09/17.

[166] Allen EA, Liu J, Kiehl KA, Gelernter J, Pearlson GD, Perrone-Bizzozero NI, et al. Components of cross-frequency modulation in health and disease. Frontiers in systems neuroscience. 2011;5:59. Epub 2011/08/03.

[167] Higashima M, Takeda T, Kikuchi M, Nagasawa T, Koshino Y. Functional connectivity between hemispheres and schizophrenic symptoms: a longitudinal study of interhemispheric EEG coherence in patients with acute exacerbations of schizophrenia. Clin EEG Neurosci. 2006;37(1):10-5. Epub 2006/02/16.

[168] Korostenskaja M, Kahkonen S. What do ERPs and ERFs reveal about the effect of antipsychotic treatment on cognition in schizophrenia? Current Pharmaceutical Design. 2009;15:(In press).

[169] Turetsky BI, Calkins ME, Light GA, Olincy A, Radant AD, Swerdlow NR. Neurophysiological endophenotypes of schizophrenia: the viability of selected candidate measures. Schizophr Bull. 2007;33(1):69-94. Epub 2006/12/01.

[170] Carrasco M, Harbin SM, Nienhuis JK, Fitzgerald KD, Gehring WJ, Hanna GL. Increased error-related brain activity in youth with obsessive-compulsive disorder and unaffected siblings. Depress Anxiety. 2013;30(1):39-46. Epub 2012/12/12.

[171] Nolfe G, Serra FP, Palma V, Buscaino GA. Brainstem involvement in obsessive-compulsive disorder. Biological psychology. 1998;48(1):69-77. Epub 1998/07/24.

[172] Sommer W. Selective attention differentially affects brainstem auditory evoked potentials of electrodermal responders and nonresponders. Psychiatry Res. 1985;16(3): 227-32. Epub 1985/11/01.

[173] Mubarak A, Badawy A. The effect of smoking on brainstem auditory evoked potentials in positive- and negative-symptom schizophrenia. CNS Spectr. 2001;6(6):514-6, 9-22. Epub 2005/03/04.

[174] Kallstrand J, Nehlstedt SF, Skold ML, Nielzen S. Lateral asymmetry and reduced forward masking effect in early brainstem auditory evoked responses in schizophrenia. Psychiatry Res. 2012;196(2-3):188-93. Epub 2012/02/14.

[175] Picton TW, John MS, Dimitrijevic A, Purcell D. Human auditory steady-state responses. Int J Audiol. 2003;42(4):177-219. Epub 2003/06/07.

[176] Plourde G. Auditory evoked potentials. Best practice & research Clinical anaesthesiology. 2006;20(1):129-39. Epub 2006/04/26.

[177] Heinrichs-Graham E, Wilson TW. Presence of strong harmonics during visual entrainment: a magnetoencephalography study. Biological psychology. 2012;91(1): 59-64. Epub 2012/05/10.

[178] Jenkins J, 3rd, Rhone AE, Idsardi WJ, Simon JZ, Poeppel D. The elicitation of audiovisual steady-state responses: multi-sensory signal congruity and phase effects. Brain Topogr. 2011;24(2):134-48. Epub 2011/03/08.

[179] Brenner CA, Krishnan GP, Vohs JL, Ahn WY, Hetrick WP, Morzorati SL, et al. Steady state responses: electrophysiological assessment of sensory function in schizophrenia. Schizophr Bull. 2009;35(6):1065-77. Epub 2009/09/04.

[180] Ferrarelli F, Massimini M, Peterson MJ, Riedner BA, Lazar M, Murphy MJ, et al. Reduced evoked gamma oscillations in the frontal cortex in schizophrenia patients: a TMS/EEG study. Am J Psychiatry. 2008;165(8):996-1005. Epub 2008/05/17.

[181] Kwon JS, O'Donnell BF, Wallenstein GV, Greene RW, Hirayasu Y, Nestor PG, et al. Gamma frequency-range abnormalities to auditory stimulation in schizophrenia. Arch Gen Psychiatry. 1999;56(11):1001-5. Epub 1999/11/24.

[182] Light GA, Hsu JL, Hsieh MH, Meyer-Gomes K, Sprock J, Swerdlow NR, et al. Gamma band oscillations reveal neural network cortical coherence dysfunction in schizophrenia patients. Biol Psychiatry. 2006;60(11):1231-40. Epub 2006/08/09.

[183] Hamm JP, Gilmore CS, Clementz BA. Augmented gamma band auditory steady-state responses: support for NMDA hypofunction in schizophrenia. Schizophr Res. 2012;138(1):1-7. Epub 2012/05/01.

[184] Griskova-Bulanova I, Dapsys K, Maciulis V, Arnfred SM. Closed eyes condition increases auditory brain responses in schizophrenia. Psychiatry Res. 2013;211(2):183-5. Epub 2012/11/15.

[185] Bolton D, Gibb W, Lees A, Raven P, Gray JA, Chen E, et al. Neurological soft signs in obsessive compulsive disorder: standardised assessment and comparison with schizophrenia. Behav Neurol. 1998;11(4):197-204. Epub 2001/09/25.

[186] Hazen EP, Reichert EL, Piacentini JC, Miguel EC, do Rosario MC, Pauls D, et al. Case series: Sensory intolerance as a primary symptom of pediatric OCD. Ann Clin Psychiatry. 2008;20(4):199-203. Epub 2008/11/27.

[187] Miguel EC, do Rosario-Campos MC, Prado HS, do Valle R, Rauch SL, Coffey BJ, et al. Sensory phenomena in obsessive-compulsive disorder and Tourette's disorder. J Clin Psychiatry. 2000;61(2):150-6; quiz 7. Epub 2000/03/25.

[188] Heckers S. Neural models of schizophrenia. Dialogues in clinical neuroscience. 2000;2(3):267-79. Epub 2000/09/01.

[189] Gauntlett-Gilbert J, Kuipers E. Visual hallucinations in psychiatric conditions: appraisals and their relationship to distress. Br J Clin Psychol. 2005;44(Pt 1):77-87. Epub 2005/04/14.

[190] Dierks T, Linden DE, Jandl M, Formisano E, Goebel R, Lanfermann H, et al. Activation of Heschl's gyrus during auditory hallucinations. Neuron. 1999;22(3):615-21. Epub 1999/04/10.

[191] Raij TT, Valkonen-Korhonen M, Holi M, Therman S, Lehtonen J, Hari R. Reality of auditory verbal hallucinations. Brain. 2009;132(Pt 11):2994-3001. Epub 2009/07/22.

[192] Kompus K, Falkenberg LE, Bless JJ, Johnsen E, Kroken RA, Krakvik B, et al. The role of the primary auditory cortex in the neural mechanism of auditory verbal hallucinations. Front Hum Neurosci. 2013;7:144. Epub 2013/05/01.

[193] Korostenskaja M, Harris E, Giovanetti C, Horn P, Wang Y, Rose D, et al. Magnetoencephalography reveals altered auditory information processing in youth with obsessive-compulsive disorder. Psychiatry Res. 2013;212(2):132-40. Epub 2013/04/03.

[194] Oades RD, Zerbin D, Dittmann-Balcar A, Eggers C. Auditory event-related potential (ERP) and difference-wave topography in schizophrenic patients with/without active hallucinations and delusions: a comparison with young obsessive-compulsive disorder (OCD) and healthy subjects. Int J Psychophysiol. 1996;22(3):185-214. Epub 1996/06/01.

[195] Morault PM, Bourgeois M, Laville J, Bensch C, Paty J. Psychophysiological and clinical value of event-related potentials in obsessive-compulsive disorder. Biol Psychiatry. 1997;42(1):46-56. Epub 1997/07/01.

[196] Kim YY, Roh AY, Yoo SY, Kang DH, Kwon JS. Impairment of source memory in patients with obsessive-compulsive disorder: equivalent current dipole analysis. Psychiatry Res. 2009;165(1-2):47-59. Epub 2008/11/26.

[197] Goncalves OF, Carvalho S, Leite J, Pocinho F, Relvas J, Fregni F. Obsessive Compulsive Disorder as a functional interhemispheric imbalance at the thalamic level. Med Hypotheses. 2011;77(3):445-7. Epub 2011/07/09.

[198] Shim G, Jung WH, Choi JS, Jung MH, Jang JH, Park JY, et al. Reduced cortical folding of the anterior cingulate cortex in obsessive-compulsive disorder. J Psychiatry Neurosci. 2009;34(6):443-9. Epub 2009/12/02.

[199] Chiu CH, Lo YC, Tang HS, Liu IC, Chiang WY, Yeh FC, et al. White matter abnormalities of fronto-striato-thalamic circuitry in obsessive-compulsive disorder: A study using diffusion spectrum imaging tractography. Psychiatry Res. 2011;192(3):176-82. Epub 2011/05/07.

[200] Szeszko PR, Christian C, Macmaster F, Lencz T, Mirza Y, Taormina SP, et al. Gray matter structural alterations in psychotropic drug-naive pediatric obsessive-compul-

sive disorder: an optimized voxel-based morphometry study. Am J Psychiatry. 2008;165(10):1299-307. Epub 2008/04/17.

[201] Perros P, Young ES, Ritson JJ, Price GW, Mann P. Power spectral EEG analysis and EEG variability in obsessive-compulsive disorder. Brain Topogr. 1992;4(3):187-92. Epub 1992/01/01.

[202] Denys D, Zohar J, Westenberg HG. The role of dopamine in obsessive-compulsive disorder: preclinical and clinical evidence. J Clin Psychiatry. 2004;65 Suppl 14:11-7. Epub 2004/11/24.

[203] van der Wee NJ, Stevens H, Hardeman JA, Mandl RC, Denys DA, van Megen HJ, et al. Enhanced dopamine transporter density in psychotropic-naive patients with obsessive-compulsive disorder shown by [123I]{beta}-CIT SPECT. Am J Psychiatry. 2004;161(12):2201-6. Epub 2004/12/01.

[204] Korostenskaja M, Kicic D, Kahkonen S. The effect of methylphenidate on auditory information processing in healthy volunteers: a combined EEG/MEG study. Psychopharmacology. 2008;197(3):475-86. Epub 2008/02/12.

[205] Catts SV, Armstrong MS, Ward PB, McConaghy N. Reduced P200 latency and allusive thinking: an auditory evoked potential index of a cognitive predisposition to schizophrenia? The International journal of neuroscience. 1986;30(3):173-9. Epub 1986/09/01.

[206] Chiarenza GA, Papakostopoulos D, Dini M, Cazzullo CL. Neurophysiological correlates of psychomotor activity in chronic schizophrenics. Electroencephalography and clinical neurophysiology. 1985;61(4):218-28. Epub 1985/10/01.

[207] Oades RD, Dittmann-Balcar A, Schepker R, Eggers C, Zerbin D. Auditory event-related potentials (ERPs) and mismatch negativity (MMN) in healthy children and those with attention-deficit or tourette/tic symptoms. Biological psychology. 1996;43(2):163-85. Epub 1996/04/12.

[208] Herrmann MJ, Jacob C, Unterecker S, Fallgatter AJ. Reduced response-inhibition in obsessive-compulsive disorder measured with topographic evoked potential mapping. Psychiatry Res. 2003;120(3):265-71. Epub 2003/10/17.

[209] Kim MS, Kim YY, Yoo SY, Kwon JS. Electrophysiological correlates of behavioral response inhibition in patients with obsessive-compulsive disorder. Depress Anxiety. 2007;24(1):22-31. Epub 2006/08/26.

[210] Savage CR, Weilburg JB, Duffy FH, Baer L, Shera DM, Jenike MA. Low-level sensory processing in obsessive-compulsive disorder: an evoked potential study. Biol Psychiatry. 1994;35(4):247-52. Epub 1994/02/15.

[211] Ciesielski KT, Beech HR, Gordon PK. Some electrophysiological observations in obsessional states. Br J Psychiatry. 1981;138:479-84. Epub 1981/06/01.

[212] Leavitt VM, Molholm S, Ritter W, Shpaner M, Foxe JJ. Auditory processing in schizophrenia during the middle latency period (10-50 ms): high-density electrical mapping and source analysis reveal subcortical antecedents to early cortical deficits. J Psychiatry Neurosci. 2007;32(5):339-53. Epub 2007/09/08.

[213] Lalor EC, De Sanctis P, Krakowski MI, Foxe JJ. Visual sensory processing deficits in schizophrenia: is there anything to the magnocellular account? Schizophr Res. 2012;139(1-3):246-52. Epub 2012/06/19.

[214] Lalor EC, Yeap S, Reilly RB, Pearlmutter BA, Foxe JJ. Dissecting the cellular contributions to early visual sensory processing deficits in schizophrenia using the VESPA evoked response. Schizophr Res. 2008;98(1-3):256-64. Epub 2007/11/13.

[215] Bedwell JS, Chan CC, Trachik BJ, Rassovsky Y. Changes in the visual-evoked P1 potential as a function of schizotypy and background color in healthy young adults. J Psychiatr Res. 2013;47(4):542-7. Epub 2013/02/02.

[216] Haenschel C, Bittner RA, Haertling F, Rotarska-Jagiela A, Maurer K, Singer W, et al. Contribution of impaired early-stage visual processing to working memory dysfunction in adolescents with schizophrenia: a study with event-related potentials and functional magnetic resonance imaging. Arch Gen Psychiatry. 2007;64(11):1229-40. Epub 2007/11/07.

[217] Koychev I, El-Deredy W, Haenschel C, Deakin JF. Visual information processing deficits as biomarkers of vulnerability to schizophrenia: an event-related potential study in schizotypy. Neuropsychologia. 2010;48(7):2205-14. Epub 2010/04/24.

[218] Oribe N, Hirano Y, Kanba S, del Re EC, Seidman LJ, Mesholam-Gately R, et al. Early and late stages of visual processing in individuals in prodromal state and first episode schizophrenia: an ERP study. Schizophr Res. 2013;146(1-3):95-102. Epub 2013/02/26.

[219] Hegerl U, Gaebel W, Ulrich G. Relationship between residual symptomatology and auditory evoked potentials in schizophrenic outpatients. Pharmacopsychiatry. 1988;21(6):329-30. Epub 1988/11/01.

[220] Schwarzkopf SB, Lamberti JS, Jiminez M, Kane CF, Henricks M, Nasrallah HA. Visual evoked potential correlates of positive/negative symptoms in schizophrenia. Biol Psychiatry. 1990;27(4):400-10. Epub 1990/02/15.

[221] Roth WT, Pfefferbaum A, Kelly AF, Berger PA, Kopell BS. Auditory event-related potentials in schizophrenia and depression. Psychiatry Res. 1981;4(2):199-212. Epub 1981/04/01.

[222] Laurent A, Garcia-Larrea L, d'Amato T, Bosson JL, Saoud M, Marie-Cardine M, et al. Auditory event-related potentials and clinical scores in unmedicated schizophrenic patients. Psychiatry Res. 1999;86(3):229-38. Epub 1999/09/11.

[223] Shenton ME, Faux SF, McCarley RW, Ballinger R, Coleman M, Torello M, et al. Correlations between abnormal auditory P300 topography and positive symptoms in

schizophrenia: a preliminary report. Biol Psychiatry. 1989;25(6):710-6. Epub 1989/03/15.

[224] Oertel-Knochel V, Linden DE. Cerebral asymmetry in schizophrenia. Neuroscientist. 2011;17(5):456-67. Epub 2011/04/27.

[225] Ribolsi M, Koch G, Magni V, Di Lorenzo G, Rubino IA, Siracusano A, et al. Abnormal brain lateralization and connectivity in schizophrenia. Rev Neurosci. 2009;20(1): 61-70. Epub 2009/06/17.

[226] Spironelli C, Angrilli A, Stegagno L. Failure of language lateralization in schizophrenia patients: an ERP study on early linguistic components. J Psychiatry Neurosci. 2008;33(3):235-43. Epub 2008/07/02.

[227] Oertel-Knochel V, Knochel C, Stablein M, Linden DE. Abnormal functional and structural asymmetry as biomarker for schizophrenia. Current topics in medicinal chemistry. 2012;12(21):2434-51. Epub 2013/01/03.

[228] Naatanen R, Gaillard AW, Mantysalo S. Early selective-attention effect on evoked potential reinterpreted. Acta psychologica. 1978;42(4):313-29.

[229] Naatanen R, Michie PT. Early selective-attention effects on the evoked potential: a critical review and reinterpretation. Biological psychology. 1979;8(2):81-136.

[230] Kujala T, Tervaniemi M, Schroger E. The mismatch negativity in cognitive and clinical neuroscience: theoretical and methodological considerations. Biol Psychol. 2007;74(1):1-19.

[231] Haenschel C, Vernon DJ, Dwivedi P, Gruzelier JH, Baldeweg T. Event-related brain potential correlates of human auditory sensory memory-trace formation. J Neurosci. 2005;25(45):10494-501.

[232] Towey JP, Tenke CE, Bruder GE, Leite P, Friedman D, Liebowitz M, et al. Brain event-related potential correlates of overfocused attention in obsessive-compulsive disorder. Psychophysiology. 1994;31(6):535-43. Epub 1994/11/01.

[233] Kahkonen S, Makinen V, Jaaskelainen IP, Pennanen S, Liesivuori J, Ahveninen J. Serotonergic modulation of mismatch negativity. Psychiatry Res. 2005;138(1):61-74.

[234] Ahveninen J, Kahkonen S, Pennanen S, Liesivuori J, Ilmoniemi RJ, Jaaskelainen IP. Tryptophan depletion effects on EEG and MEG responses suggest serotonergic modulation of auditory involuntary attention in humans. NeuroImage. 2002;16(4): 1052-61. Epub 2002/08/31.

[235] Heekeren K, Daumann J, Neukirch A, Stock C, Kawohl W, Norra C, et al. Mismatch negativity generation in the human 5HT2A agonist and NMDA antagonist model of psychosis. Psychopharmacology. 2008;199(1):77-88. Epub 2008/05/20.

[236] Umbricht D, Krljes S. Mismatch negativity in schizophrenia: a meta-analysis. Schizophr Res. 2005;76(1):1-23.

[237] Umbricht DS, Bates JA, Lieberman JA, Kane JM, Javitt DC. Electrophysiological indices of automatic and controlled auditory information processing in first-episode, recent-onset and chronic schizophrenia. Biol Psychiatry. 2006;59(8):762-72.

[238] Shelley AM, Ward PB, Catts SV, Michie PT, Andrews S, McConaghy N. Mismatch negativity: an index of a preattentive processing deficit in schizophrenia. Biol Psychiatry. 1991;30(10):1059-62.

[239] Michie PT. What has MMN revealed about the auditory system in schizophrenia? Int J Psychophysiol. 2001;42(2):177-94.

[240] Javitt DC. Intracortical mechanisms of mismatch negativity dysfunction in schizophrenia. Audiol Neurootol. 2000;5(3-4):207-15.

[241] Kreitschmann-Andermahr I, Rosburg T, Meier T, Volz HP, Nowak H, Sauer H. Impaired sensory processing in male patients with schizophrenia: a magnetoencephalographic study of auditory mismatch detection. Schizophr Res. 1999;35(2):121-9.

[242] Hirayasu Y, Potts GF, O'Donnell BF, Kwon JS, Arakaki H, Akdag SJ, et al. Auditory mismatch negativity in schizophrenia: topographic evaluation with a high-density recording montage. Am J Psychiatry. 1998;155(9):1281-4.

[243] Hirayasu Y, Shenton ME, Salisbury DF, Dickey CC, Fischer IA, Mazzoni P, et al. Lower left temporal lobe MRI volumes in patients with first-episode schizophrenia compared with psychotic patients with first-episode affective disorder and normal subjects. Am J Psychiatry. 1998;155(10):1384-91.

[244] Wolf RC, Hose A, Frasch K, Walter H, Vasic N. Volumetric abnormalities associated with cognitive deficits in patients with schizophrenia. Eur Psychiatry. 2008. Epub 2008/04/25.

[245] Salisbury DF, Kuroki N, Kasai K, Shenton ME, McCarley RW. Progressive and interrelated functional and structural evidence of post-onset brain reduction in schizophrenia. Arch Gen Psychiatry. 2007;64(5):521-9.

[246] Catts SV, Shelley AM, Ward PB, Liebert B, McConaghy N, Andrews S, et al. Brain potential evidence for an auditory sensory memory deficit in schizophrenia. Am J Psychiatry. 1995;152(2):213-9.

[247] Fujimoto T, Okumura E, Takeuchi K, Kodabashi A, Otsubo T, Nakamura K, et al. Dysfunctional cortical connectivity during the auditory oddball task in patients with schizophrenia. The open neuroimaging journal. 2013;7:15-26. Epub 2013/06/12.

[248] Korostenskaja M, Nikulin VV, Kicic D, Nikulina AV, Kahkonen S. Effects of NMDA receptor antagonist memantine on mismatch negativity. Brain research bulletin. 2007;72(4-6):275-83.

[249] Kreitschmann-Andermahr I, Rosburg T, Demme U, Gaser E, Nowak H, Sauer H. Effect of ketamine on the neuromagnetic mismatch field in healthy humans. Brain Res Cogn Brain Res. 2001;12(1):109-16.

[250] Rosburg T, Marinou V, Haueisen J, Smesny S, Sauer H. Effects of lorazepam on the neuromagnetic mismatch negativity (MMNm) and auditory evoked field component N100m. Neuropsychopharmacology. 2004;29(9):1723-33. Epub 2004/05/06.

[251] Korostenskaja M, Pekkonen E, Horn P, Huttunen J, Kivisaari R, Kahkonen S. Mismatch negativity (MMN) reflects changes in GABA system: combined EEG/MEG study. EPIC XV 15th International Congress on Event-Related Potentials of the Brain; APRIL 22 – 25, 2009; Bloomington, Indiana, USA2009. p. 50.

[252] Oades RD, Dittmann-Balcar A, Zerbin D, Grzella I. Impaired attention-dependent augmentation of MMN in nonparanoid vs paranoid schizophrenic patients: a comparison with obsessive-compulsive disorder and healthy subjects. Biol Psychiatry. 1997;41(12):1196-210. Epub 1997/06/15.

[253] Schroger E, Wolff C. Behavioral and electrophysiological effects of task-irrelevant sound change: a new distraction paradigm. Brain Res Cogn Brain Res. 1998;7(1): 71-87.

[254] Escera C, Yago E, Alho K. Electrical responses reveal the temporal dynamics of brain events during involuntary attention switching. Eur J Neurosci. 2001;14(5):877-83.

[255] Friedman D, Cycowicz YM, Gaeta H. The novelty P3: an event-related brain potential (ERP) sign of the brain's evaluation of novelty. Neurosci Biobehav Rev. 2001;25(4): 355-73. Epub 2001/07/11.

[256] Knight RT. Decreased response to novel stimuli after prefrontal lesions in man. Electroencephalography and clinical neurophysiology. 1984;59(1):9-20. Epub 1984/02/01.

[257] Alho K, Winkler I, Escera C, Huotilainen M, Virtanen J, Jaaskelainen IP, et al. Processing of novel sounds and frequency changes in the human auditory cortex: magnetoencephalographic recordings. Psychophysiology. 1998;35(2):211-24. Epub 1998/04/08.

[258] Kahkonen S, Marttinen Rossi E, Yamashita H. Alcohol impairs auditory processing of frequency changes and novel sounds: a combined MEG and EEG study. Psychopharmacology. 2005;177(4):366-72. Epub 2004/08/04.

[259] Sutton S, Braren M, Zubin J, John ER. Evoked-potential correlates of stimulus uncertainty. Science. 1965;150(700):1187-8.

[260] Polich J. Clinical application of the P300 event-related brain potential. Phys Med Rehabil Clin N Am. 2004;15(1):133-61.

[261] Squires NK, Squires KC, Hillyard SA. Two varieties of long-latency positive waves evoked by unpredictable auditory stimuli in man. Electroencephalography and clinical neurophysiology. 1975;38(4):387-401.

[262] Karakas S, Basar E. Models and theories of brain function in cognition within a framework of behavioral cognitive psychology. Int J Psychophysiol. 2006;60(2): 186-93.

[263] Magliero A, Bashore TR, Coles MG, Donchin E. On the dependence of P300 latency on stimulus evaluation processes. Psychophysiology. 1984;21(2):171-86.

[264] Polich J, Howard L, Starr A. Effects of age on the P300 component of the event-related potential from auditory stimuli: peak definition, variation, and measurement. J Gerontol. 1985;40(6):721-6.

[265] Ischebeck M, Endrass T, Simon D, Kathmann N. Auditory novelty processing is enhanced in obsessive-compulsive disorder. Depress Anxiety. 2011;28(10):915-23. Epub 2011/09/08.

[266] Gohle D, Juckel G, Mavrogiorgou P, Pogarell O, Mulert C, Rujescu D, et al. Electrophysiological evidence for cortical abnormalities in obsessive-compulsive disorder - a replication study using auditory event-related P300 subcomponents. J Psychiatr Res. 2008;42(4):297-303. Epub 2007/03/03.

[267] Mavrogiorgou P, Juckel G, Frodl T, Gallinat J, Hauke W, Zaudig M, et al. P300 subcomponents in obsessive-compulsive disorder. J Psychiatr Res. 2002;36(6):399-406. Epub 2002/10/24.

[268] Beech HR, Ciesielski KT, Gordon PK. Further observations of evoked potentials in obsessional patients. Br J Psychiatry. 1983;142:605-9. Epub 1983/06/01.

[269] Towey J, Bruder G, Hollander E, Friedman D, Erhan H, Liebowitz M, et al. Endogenous event-related potentials in obsessive-compulsive disorder. Biol Psychiatry. 1990;28(2):92-8. Epub 1990/07/15.

[270] Miyata A, Matsunaga H, Kiriike N, Iwasaki Y, Takei Y, Yamagami S. Event-related potentials in patients with obsessive-compulsive disorder. Psychiatry Clin Neurosci. 1998;52(5):513-8. Epub 1999/04/24.

[271] Atkinson RJ, Michie PT, Schall U. Duration mismatch negativity and p3a in first-episode psychosis and individuals at ultra-high risk of psychosis. Biol Psychiatry. 2012;71(2):98-104. Epub 2011/10/18.

[272] Jahshan C, Cadenhead KS, Rissling AJ, Kirihara K, Braff DL, Light GA. Automatic sensory information processing abnormalities across the illness course of schizophrenia. Psychol Med. 2012;42(1):85-97. Epub 2011/07/12.

[273] Kaur M, Battisti RA, Ward PB, Ahmed A, Hickie IB, Hermens DF. MMN/P3a deficits in first episode psychosis: comparing schizophrenia-spectrum and affective-spectrum subgroups. Schizophr Res. 2011;130(1-3):203-9. Epub 2011/05/10.

[274] Gil-da-Costa R, Stoner GR, Fung R, Albright TD. Nonhuman primate model of schizophrenia using a noninvasive EEG method. Proc Natl Acad Sci U S A. 2013. Epub 2013/08/21.

[275] Jahshan C, Wynn JK, Mathis KI, Altshuler LL, Glahn DC, Green MF. Cross-diagnostic comparison of duration mismatch negativity and P3a in bipolar disorder and schizophrenia. Bipolar Disord. 2012;14(3):239-48. Epub 2012/05/03.

[276] van der Stelt O, Belger A. Application of electroencephalography to the study of cognitive and brain functions in schizophrenia. Schizophr Bull. 2007;33(4):955-70. Epub 2007/03/17.

[277] Ford JM, Sullivan EV, Marsh L, White PM, Lim KO, Pfefferbaum A. The relationship between P300 amplitude and regional gray matter volumes depends upon the attentional system engaged. Electroencephalography and clinical neurophysiology. 1994;90(3):214-28.

[278] Juckel G, Muller-Schubert A, Gaebel W, Hegerl U. Residual symptoms and P300 in schizophrenic outpatients. Psychiatry Res. 1996;65(1):23-32.

[279] Molina V, Sanz J, Munoz F, Casado P, Hinojosa JA, Sarramea F, et al. Dorsolateral prefrontal cortex contribution to abnormalities of the P300 component of the event-related potential in schizophrenia. Psychiatry Res. 2005;140(1):17-26.

[280] Roth WT, Cannon EH. Some features of the auditory evoked response in schizophrenics. Arch Gen Psychiatry. 1972;27(4):466-71.

[281] Hirayasu Y. Brain imaging in schizophrenia. Neuropathology. 2007;27(6):601-3.

[282] Ford JM, Pfefferbaum A, Roth W. P3 and schizophrenia. Ann N Y Acad Sci. 1992;658:146-62.

[283] Sumich A, Kumari V, Dodd P, Ettinger U, Hughes C, Zachariah E, et al. N100 and P300 amplitude to Go and No-Go variants of the auditory oddball in siblings discordant for schizophrenia. Schizophr Res. 2008;98(1-3):265-77.

[284] Kawasaki Y, Sumiyoshi T, Higuchi Y, Ito T, Takeuchi M, Kurachi M. Voxel-based analysis of P300 electrophysiological topography associated with positive and negative symptoms of schizophrenia. Schizophr Res. 2007;94(1-3):164-71.

[285] Pfefferbaum A, Wenegrat BG, Ford JM, Roth WT, Kopell BS. Clinical application of the P3 component of event-related potentials. II. Dementia, depression and schizophrenia. Electroencephalography and clinical neurophysiology. 1984;59(2):104-24.

[286] Araki T, Kasai K, Kirihara K, Yamasue H, Kato N, Kudo N, et al. Auditory P300 latency prolongation with age in schizophrenia: gender and subcomponent effects. Schizophr Res. 2006;88(1-3):217-21.

[287] Mori Y, Kurosu S, Hiroyama Y, Niwa S. Prolongation of P300 latency is associated with the duration of illness in male schizophrenia patients. Psychiatry Clin Neurosci. 2007;61(5):471-8.

[288] Nieman DH, Koelman JH, Linszen DH, Bour LJ, Dingemans PM, Ongerboer de Visser BW. Clinical and neuropsychological correlates of the P300 in schizophrenia. Schizophr Res. 2002;55(1-2):105-13.

[289] Kim DW, Shim M, Kim JI, Im CH, Lee SH. Source Activation of P300 Correlates with Negative Symptom Severity in Patients with Schizophrenia. Brain Topogr. 2013. Epub 2013/07/31.

[290] Kim MS, Kang SS, Youn T, Kang DH, Kim JJ, Kwon JS. Neuropsychological correlates of P300 abnormalities in patients with schizophrenia and obsessive-compulsive disorder. Psychiatry Res. 2003;123(2):109-23. Epub 2003/07/10.

[291] Pallanti S, Castellini G, Chamberlain SR, Quercioli L, Zaccara G, Fineberg NA. Cognitive event-related potentials differentiate schizophrenia with obsessive-compulsive disorder (schizo-OCD) from OCD and schizophrenia without OC symptoms. Psychiatry Res. 2009;170(1):52-60. Epub 2009/10/06.

[292] Gehring WJ, Himle J, Nisenson LG. Action-monitoring dysfunction in obsessive-compulsive disorder. Psychological science. 2000;11(1):1-6. Epub 2001/03/07.

[293] Falkenstein M, Hohnsbein J, Hoormann J, Blanke L. Effects of crossmodal divided attention on late ERP components. II. Error processing in choice reaction tasks. Electroencephalography and clinical neurophysiology. 1991;78(6):447-55. Epub 1991/06/01.

[294] Falkenstein M, Hoormann J, Christ S, Hohnsbein J. ERP components on reaction errors and their functional significance: a tutorial. Biological psychology. 2000;51(2-3): 87-107. Epub 2000/02/25.

[295] Endrass T, Klawohn J, Schuster F, Kathmann N. Overactive performance monitoring in obsessive-compulsive disorder: ERP evidence from correct and erroneous reactions. Neuropsychologia. 2008;46(7):1877-87. Epub 2008/06/03.

[296] Grundler TO, Cavanagh JF, Figueroa CM, Frank MJ, Allen JJ. Task-related dissociation in ERN amplitude as a function of obsessive-compulsive symptoms. Neuropsychologia. 2009;47(8-9):1978-87. Epub 2009/05/12.

[297] Hajcak G, Franklin ME, Foa EB, Simons RF. Increased error-related brain activity in pediatric obsessive-compulsive disorder before and after treatment. Am J Psychiatry. 2008;165(1):116-23. Epub 2007/11/08.

[298] Johannes S, Wieringa BM, Nager W, Rada D, Dengler R, Emrich HM, et al. Discrepant target detection and action monitoring in obsessive-compulsive disorder. Psychiatry Res. 2001;108(2):101-10. Epub 2001/12/12.

[299] Ruchsow M, Spitzer M, Gron G, Grothe J, Kiefer M. Error processing and impulsiveness in normals: evidence from event-related potentials. Brain Res Cogn Brain Res. 2005;24(2):317-25. Epub 2005/07/05.

[300] Santesso DL, Segalowitz SJ, Schmidt LA. Error-related electrocortical responses are enhanced in children with obsessive-compulsive behaviors. Dev Neuropsychol. 2006;29(3):431-45. Epub 2006/05/05.

[301] Nieuwenhuis S, Nielen MM, Mol N, Hajcak G, Veltman DJ. Performance monitoring in obsessive-compulsive disorder. Psychiatry Res. 2005;134(2):111-22. Epub 2005/04/21.

[302] Endrass T, Schuermann B, Kaufmann C, Spielberg R, Kniesche R, Kathmann N. Performance monitoring and error significance in patients with obsessive-compulsive disorder. Biological psychology. 2010;84(2):257-63. Epub 2010/02/16.

[303] Bates AT, Kiehl KA, Laurens KR, Liddle PF. Error-related negativity and correct response negativity in schizophrenia. Clin Neurophysiol. 2002;113(9):1454-63. Epub 2002/08/10.

[304] Kopp B, Rist F. An event-related brain potential substrate of disturbed response monitoring in paranoid schizophrenic patients. J Abnorm Psychol. 1999;108(2): 337-46. Epub 1999/06/16.

[305] Morris SE, Yee CM, Nuechterlein KH. Electrophysiological analysis of error monitoring in schizophrenia. J Abnorm Psychol. 2006;115(2):239-50. Epub 2006/06/02.

[306] Foti D, Kotov R, Bromet E, Hajcak G. Beyond the broken error-related negativity: functional and diagnostic correlates of error processing in psychosis. Biol Psychiatry. 2012;71(10):864-72. Epub 2012/02/18.

[307] Millet B, Kochman F, Gallarda T, Krebs MO, Demonfaucon F, Barrot I, et al. Phenomenological and comorbid features associated in obsessive-compulsive disorder: influence of age of onset. J Affect Disord. 2004;79(1-3):241-6. Epub 2004/03/17.

[308] JAMA. Cognitive-behavior therapy, sertraline, and their combination for children and adolescents with obsessive-compulsive disorder: the Pediatric OCD Treatment Study (POTS) randomized controlled trial. JAMA. 2004;292(16):1969-76. Epub 2004/10/28.

[309] Insel TR, Mueller EA, 3rd, Gillin JC, Siever LJ, Murphy DL. Biological markers in obsessive-compulsive and affective disorders. J Psychiatr Res. 1984;18(4):407-23. Epub 1984/01/01.

[310] Khodayari-Rostamabad A, Hasey GM, Maccrimmon DJ, Reilly JP, de Bruin H. A pilot study to determine whether machine learning methodologies using pre-treatment

electroencephalography can predict the symptomatic response to clozapine therapy. Clin Neurophysiol. 2010;121(12):1998-2006. Epub 2010/11/03.

[311] Ralescu A, Lee KH, Korostenskaja M. Machine learning techniques provide with a new insight into pre-attentive information processing changes in pediatric intractable epilepsy. Society for Psychophysiological Research 51st Annual Meeting 2011; September 14-18, 2011; Boston, MA, USA2011. p. 92.

[312] Armananzas R, Alonso-Nanclares L, Defelipe-Oroquieta J, Kastanauskaite A, de Sola RG, Defelipe J, et al. Machine learning approach for the outcome prediction of temporal lobe epilepsy surgery. PLoS ONE. 2013;8(4):e62819. Epub 2013/05/07.

[313] Chang NF, Chen TC, Chiang CY, Chen LG. Channel selection for epilepsy seizure prediction method based on machine learning. Conf Proc IEEE Eng Med Biol Soc. 2012;2012:5162-5. Epub 2013/02/01.

[314] Visa S, Ralescu A. Data-driven fuzzy sets for classification. International Journal of Advanced Intelligence Paradigms. 2008;1(1):3-30.

[315] Takahashi T, Cho RY, Mizuno T, Kikuchi M, Murata T, Takahashi K, et al. Antipsychotics reverse abnormal EEG complexity in drug-naive schizophrenia: a multiscale entropy analysis. NeuroImage. 2010;51(1):173-82. Epub 2010/02/13.

[316] Javitt DC. Glutamatergic theories of schizophrenia. Isr J Psychiatry Relat Sci. 2010;47(1):4-16. Epub 2010/08/06.

[317] Nakao T, Nakagawa A, Yoshiura T, Nakatani E, Nabeyama M, Yoshizato C, et al. Brain activation of patients with obsessive-compulsive disorder during neuropsychological and symptom provocation tasks before and after symptom improvement: a functional magnetic resonance imaging study. Biol Psychiatry. 2005;57(8):901-10. Epub 2005/04/12.

[318] Rauch SL, Shin LM, Dougherty DD, Alpert NM, Fischman AJ, Jenike MA. Predictors of fluvoxamine response in contamination-related obsessive compulsive disorder: a PET symptom provocation study. Neuropsychopharmacology. 2002;27(5):782-91. Epub 2002/11/15.

[319] Saxena S, Brody AL, Maidment KM, Dunkin JJ, Colgan M, Alborzian S, et al. Localized orbitofrontal and subcortical metabolic changes and predictors of response to paroxetine treatment in obsessive-compulsive disorder. Neuropsychopharmacology. 1999;21(6):683-93. Epub 2000/01/14.

[320] Koprivova J, Congedo M, Raszka M, Prasko J, Brunovsky M, Horacek J. Prediction of treatment response and the effect of independent component neurofeedback in obsessive-compulsive disorder: a randomized, sham-controlled, double-blind study. Neuropsychobiology. 2013;67(4):210-23. Epub 2013/05/03.

[321] Basar E. The theory of the whole-brain-work. Int J Psychophysiol. 2006;60(2):133-8. Epub 2006/03/28.

[322] Robinson DL. How brain arousal systems determine different temperament types and the major dimensions of personality. Personality and Individual Differences. 2001;31:1233–59.

[323] Miskovic V, Moscovitch DA, Santesso DL, McCabe RE, Antony MM, Schmidt LA. Changes in EEG cross-frequency coupling during cognitive behavioral therapy for social anxiety disorder. Psychological science. 2011;22(4):507-16. Epub 2011/03/08.

[324] Shagass C, Roemer RA, Straumanis JJ, Josiassen RC. Distinctive somatosensory evoked potential features in obsessive-compulsive disorder. Biol Psychiatry. 1984;19(11):1507-24. Epub 1984/11/01.

[325] Shagass C, Roemer RA, Straumanis JJ, Josiassen RC. Evoked potentials in obsessive-compulsive disorder. Advances in Biological Psychiatry. 1984:1569-75.

[326] Khanna S, Mukundan CR, Channabasavanna SM. Middle latency evoked potentials in obsessive-compulsive disorder. Biol Psychiatry. 1989;25(7):980-3. Epub 1989/04/01.

[327] Towey J, Bruder G, Tenke C, Leite P, DeCaria C, Friedman D, et al. Event-related potential and clinical correlates of neurodysfunction in obsessive-compulsive disorder. Psychiatry Res. 1993;49(2):167-81. Epub 1993/11/01.

[328] Baribeau-Braun J, Picton TW, Gosselin JY. Schizophrenia: a neurophysiological evaluation of abnormal information processing. Science. 1983;219(4586):874-6. Epub 1983/02/18.

[329] de Groot CM, Torello MW, Boutros NN, Allen R. Auditory event-related potentials and statistical probability mapping in obsessive-compulsive disorder. Clin Electroencephalogr. 1997;28(3):148-54. Epub 1997/07/01.

[330] Okasha A, Rafaat M, Mahallawy N, El Nahas G, El Dawla AS, Sayed M, et al. Cognitive dysfunction in obsessive-compulsive disorder. Acta Psychiatr Scand. 2000;101(4): 281-5. Epub 2000/04/27.

[331] Kim YY, Yoo SY, Kim MS, Kwon JS. Equivalent current dipole of word repetition effects in patients with obsessive-compulsive disorder. Brain Topogr. 2006;18(3):201-12. Epub 2006/03/18.

[332] Hajcak G, Simons RF. Error-related brain activity in obsessive-compulsive undergraduates. Psychiatry Res. 2002;110(1):63-72. Epub 2002/05/15.

[333] Ruchsow M, Grön G, Reuter K, Spitzer M, Hermle L, Kiefer M. Error-Related Brain Activity in Patients with Obsessive- Compulsive Disorder and in Healthy Controls. Journal of Psychophysiology 2005;19(4):298-304.

[334] Xiao Z, Wang J, Zhang M, Li H, Tang Y, Wang Y, et al. Error-related negativity abnormalities in generalized anxiety disorder and obsessive-compulsive disorder. Progress in neuro-psychopharmacology & biological psychiatry. 2011;35(1):265-72. Epub 2010/11/30.

[335] Hanna GL, Carrasco M, Harbin SM, Nienhuis JK, LaRosa CE, Chen P, et al. Error-related negativity and tic history in pediatric obsessive-compulsive disorder. J Am Acad Child Adolesc Psychiatry. 2012;51(9):902-10. Epub 2012/08/25.

[336] Carrasco M, Hong C, Nienhuis JK, Harbin SM, Fitzgerald KD, Gehring WJ, et al. Increased error-related brain activity in youth with obsessive-compulsive disorder and other anxiety disorders. Neuroscience letters. 2013;541:214-8. Epub 2013/03/05.

Chapter 4

Effects of Obsessive-Compulsive Disorder Symptom Intensity on Brain Electro-Cortical Activity Associated with Emotional Memory

Marc E. Lavoie, Geneviève Sauvé,
Simon Morand-Beaulieu, Marie-Pierre Charron and
Kieron P. O'Connor

Additional information is available at the end of the chapter

http://dx.doi.org/10.5772/57179

1. Introduction

Obsessive-Compulsive Disorder (OCD) consists of recurrent and persistent thoughts (**obsessions**), accompanied by the development of ritual and/or repetitive behaviors (**compulsions**). It has been widely suggested that these symptoms might be linked to different types of cognitive and cerebral impairments. More precisely, it has been shown that OCD patients present specific memory difficulties that could be related to a Cortico-Striato-Thalamo-Cortical (CSTC) loop dysfunction. Some research has further explained these memory problems as a result of poor confidence, along with affected emotional evaluation.

Numerous studies have undertaken the task to examine which specific aspect of memory processing might be affected in OCD patients. For instance, the California Verbal Learning Test (CVLT) was employed to assess verbal memory. Overall, findings have been mixed with poorer encoding, but intact delayed recall, impaired encoding and long delay recall, or completely intact encoding, recall and recognition memory. In contrast, there is a stronger support for the presence of a visual memory deficit in OCD. Several studies have reported impaired recall on the Rey Complex Figure Test (RCFT), as well as recall on other tests of non verbal memory.

In parallel, a general questioning regarding a possible relationship between emotion processing and OCD symptoms remains unresolved. From a psychological measurement standpoint, obsessive and compulsive symptom expressions could also be modulated by internal mediating factors like the emotional context, including the type of material presented, its emotional valence and its complexity. These examples of internal factors could likely influence performance behavior including the degree of avoidance, hesitation and difficulties in recall and memory. Thus, a potential candidate explanation for OCD patients' altered neuropsychological performances, may lie in emotionality and emotional stressors. More precisely, stimuli eliciting anxious over-concern, strong emotionality such as those perceived as threatening, may activate a differential performance. Some investigations have focused their attention on the relationship between emotion and cognition in order to explain and find what triggers OCD behaviors. For instance, OCD patients attended to negative words longer than controls in a modified Stroop test using neutral and emotional threatening word sets. It was proposed that OCD patients have a propensity to encode negative words. In the same line, another research group found that OCD patients exhibited deficits in the ability to forget negative material compared to controls. Concurrently, others found that OCD washers showed longer latency responses to contamination words than non-washers, but that in general, OCD participants showed a longer latency to threat words. Thereby, these findings suggest that a threat-related hypothesis might explain OCD's maladaptive behaviors rather than an anxious concern and emotionality hypothesis.

In addition, other studies have indicated that the apparent memory-deficit in OCD might be modulated by the emotional valence of the presented stimulus. assessed memory in people with OCD with fear for contamination and compared their memory capacities to healthy controls. They mainly found that the OCD group had better memory skills for contaminated objects than for clean ones. Furthermore, neuropsychological tests scores indicated that this bias was not the result of differences in general memory ability. However, another investigation showed that physiological emotional reactions to stimuli were appropriate in OCD, but that patients' facial expressions indicated more attempts to suppress emotion and fear. In other words, the OCD group reacted to their emotionality more censoriously. In certain context, superior memory performances can be found with OCD checkers but they were more dissatisfied with the vividness of their recall, than were the controls. This might constitute an important clue underlying OCD's difficulties in memory confidence. Additionally, recent research suggests that there is a positive memory bias for threatening information in compulsive cleaners. However, it was proposed that the relationship between OCD and memory is likely to be more complex when the compulsive behavior is checking in a context of high responsibility. explored this relationship in a clinical sample of compulsive checkers and found a positive memory bias for threat-relevant information. In this laboratory setting, when the responsibility was artificially inflated, they observed that this positive memory bias was concurrently amplified. As expected, under conditions of no responsibility, no memory bias was detectable. However, responsibility appeared to have had a greater impact on confidence in memory than on memory itself in OCD.

In sum, the memory deficit seems to be mediated by emotionality, the degree of investment in motivations and by the responsibility to threat-relevant stimuli. However, the link between memory and emotional dimensions were not investigated yet in the context of cognitive electrophysiology and Event-Related brain potentials (ERP), which might bring some essential insights regarding cognitive processes.

2. Event-related brain potentials in obsessive-compulsive disorder

The ERP technique represents the stimulus-locked average of the raw electroencephalographic (EEG) signals. A core feature of the ERP approach is its sensitivity to covert information processing, which may not be fully assessed with behavioral measures. By giving a particular instruction during recordings, such as categorizing a stimulus by a motor response, various ERP components are obtained. These are related to different levels of cognitive processes as well as early sensory processing. Early ERP components (P100, N100, N200) are predominantly related to basic cognitive functions such as selective attention and mismatch detection, whereas the middle and late components (P300) vary with higher order information processing, such as stimulus evaluation and working memory storage.

Several ERP studies hypothesized over-focused attention in OCD, as manifested by shorter latency P300, particularly when task difficulty was increased. It has also been reported that during attention tasks, OCD was associated with larger processing negativity over the frontal area and a smaller P300 to attended target-stimuli, while the N200 was intact. Similar findings, specific to OCD patients, were reported with a visual Oddball task showing reduced P300 amplitude to target-stimuli particularly in the anterior region. Furthermore, specifically to OCD, while the P300 component appeared earlier, both N100 and P200 components oppositely showed a delayed latency. Together, these ERP findings suggest that in OCD, there is a misallocation of cognitive resources, particularly regarding memory updating and stimulus evaluation timing. However, no experiment was designed to assess episodic memory, in the context of emotional modulation.

3. Goal and hypotheses

Neuropsychological studies of OCD have most often reported minimal or no significant correlations between OCD symptom severity and cognitive functioning in general. Concurrently, some have effectively observed relationships, but more specific to certain memory or executive function tests. We propose that the discrepancy in past results could reside in part in the sub-optimal selection of OCD patients without taking into account the severity and the comorbid symptoms such as depression and anxiety. Thus, the current project mainly aimed to investigate the impact of OCD symptom severity and comorbidity on emotional memory and ERPs. Since the emotional influence on memory for OCD patients has never been explored with electrophysiological measures, our results might bring a more precise understanding of the phenomenon, especially regarding the temporal resolution of these memory stages. The use of pictures instead of words and separating the two groups by their symptom intensity constitute original ways to analyze memory processes in OCD. We also intend to assess subjective memory confidence concurrently. Based on earlier studies of OCD participants, we predicted that emotions will impact on memory. The classical episodic memory (EM) effect will be less important in both patient groups than for the control group. In addition, severe OCD patients will show the most important impairments.

4. Methodology

4.1. Participants

Two OCD groups were separated based on their symptom severity (median split at 29) with the Yale-Brown Obsessive Compulsive Scale (Y-BOCS; Goodman, Price, Rasmussen, Mazure, Delgado, et al., 1989). Fourteen extreme OCD (OCD+) were compared to 15 less symptomatic OCD (OCD-), matched to 15 controls on age, sex, laterality, mother tongue and education (see table 1).

	OCD + (N=14)	OCD - (N=15)	CTRL (N=15)	sig
Age (Years)	38 (10)	39 (10)	35 (10)	ns
Sex (F/M)	9/5	8/8	10/5	ns
Laterality (L/R/A)	1/9/2	0/11/0	0/15/0	ns
Mother tongue (French/English/Other)	10/1/0	10/3/0	14/0/1	ns
Schooling (Years)	14	14	18	ns

Note. n: non significant OCD+: Obsessive-Compulsive highly symptomatic; OCD-: Obsessive-Compulsive low symptomatic; CTRL: non psychiatric control group.

Table 1. Mean and standard deviations (in parentheses) of demographic data

4.2. Clinical assessment

All groups completed the Beck Depression Inventory (BDI), the Beck Anxiety Inventory (BAI), the Y-BOCS and the Padua Inventory for OCD symptoms. The OC symptoms severity was evaluated with the self-rated Y-BOCS for all groups, while both self and clinician-rated Y-BOCS was administered in both OCD group. With our sample of OCD, the reliability between self and clinician-rated global scores were good (α = 0.71) with no statistical differences between both evaluations (p= .11). Only the self-rated Y-BOCS will be reported in the current study. Other studies confirm the validity and reliability of the scales (internal consistency = 0.91-0.94, r = 0.90). The self-rated Padua inventory was also administered to all groups and constitute a 60-item inventory of obsessions and compulsions. The total scale (α = 0.95) and the subscales (α = 0.75-0.91) are also reliable. The BAI was also administered and consist in a 21-item anxiety symptom checklist rating symptom intensity for the last week on a 0–3 scale

($\alpha = 0.91$). To assess the presence of depression, the BDI was used, which consist in a 21-item questionnaire relative to depression ($\alpha = 0.91$).

Our groups were significantly different on the Y-BOCS global score [$F(2,42) = 496.99$, $p < 0.001$] and the Padua inventory ([$F(2,42) = 11.18$, $p < 0.001$]. A post hoc Tukey revealed that the OCD + groups were different from the OCD- (p= .001) and the controls (p= .001). But OCD+ and OCD- were not statistically different on the Padua inventory global scale and three subscales (checking, precision and rumination). There was also significant differences on anxiety [$F(2,42) = 5.16$, $p < 0.01$] and depression [$F(2,42) = 14.58$, $p < 0.001$], revealing that both OCD groups were significantly different from the controls on these assessments (Table 2). Given the fact that significant scores were found in anxiety and depression, we will use these traits as covariates in the statistical analysis involving ERPs in order to partial out the effect of these comorbid variables (see statistical analysis).

	OCD + (N=14)	OCD - (N=15)	CTRL (N=15)		
Clinical scales	**A**	**B**	**C**	**ANOVA**	**Tukey**
Y-BOCS global scale	33 (3)	24 (4)	0.60 (1)	***	A > B > C
Obsessions	16 (2)	11 (4)	0.20 (1)	***	A > B > C
Compulsions	17 (2)	13 (3)	0.40 (1)	***	A > B > C
Padua Inventory	85 (60)	65 (28)	20 (13)	***	A >C B > C
Checking	19 (13)	14 (8)	3 (3)	***	A > C B > C
Precision	10 (7)	6 (4)	1 (1)	***	A > C B > C
Rumination	32 (22)	27 (11)	7 (6)	***	A > C B > C
Washing	15 (16)	15 (9)	7 (5)	ns	...
Beck Anxiety	17 (17)	14 (10)	4 (3)	***	A > C B > C
Beck Depression	22 (16)	17 (6)	3 (3)	***	A > C B > C

Note. ***: $p < 0.001$ Y-BOCS: Yale-Brown Obsessive-Compulsive Scale. OCD+: Obsessive-Compulsive high symptomatic; OCD-: Obsessive-Compulsive low symptomatic; CTRL: non psychiatric control group. ns: non significant.

Table 2. Add a space or tabulation before Checking, Precision, Rumination and Washing. They are subscales of the Padua Inventory.

4.3. Stimuli selection

The emotional materials were constituted by photographic images from the International Affective Picture Systems, a standardized collection of images gathered from a wide variety of emotional and semantic categories. A total of 150 photographic images were chosen and classified into three groups, based on the arousal and valence estimation from the IAPS normalization (50 unpleasants, 50 neutrals and 50 pleasants). The stimuli for the study phase included a total of 75 images. For the test phase, the list included the 75 images of the study phase (old), plus 75 images that had not been presented (new) before. The selected images were classified into three basic categories based on the IAPS female ratings of valence [unpleasant = 1-3; neutral = 4-6; pleasant = 7-9]. These 150 images [25 trials by 2 response types (old/new) by 3 valence categories (pleasant/unpleasant/neutral)] were presented in different orders to counterbalance potential effects due to sequence. In addition, for half of participants, the old/new order of presentation was inverted. There were no significant differences between old and new categories across valence or arousal values (all p's > .36). In each emotional category, the images contained the same basic attributes (scenes including humans, animals, inanimate objects or landscapes) across old and new category in order to preserve coherence across recall conditions.

The images were presented one at a time on a 17" SVGA monitor (Viewsonic), for a fixed duration of 4000 ms, at a distance of 90 cm calculated from the nose to the center of the computer screen with a 5 degrees angle. They were presented at a resolution of 640 x 480 pixels in 256 colors. The inter-trial interval (ITI) was fixed at 2000 ms during which a red and white checkerboard image appeared (IAPS #7182). This red and white checkerboard image informed the participant to fixate their gaze on a point between picture presentations and reduce the eye movements. This procedure also helped to reduce the after image effect, which occurred during presentation of a blank background in our previous pilots.

4.4. Experimental procedure

The experimental session began with a **study phase** during which the participants were instructed first to fix their gaze on a red and white checkerboard screen while waiting for the next images to appear. At that point, participants were told that a series of images would be presented and that they should attend to each picture the entire time it appeared on the screen without giving any response. A short retention interval of 10 minutes was allowed between the study and the test phase. In the **test phase**, images were projected for the same duration and ITI as for the study phase. The participants were instructed to detect the images that had been already presented (*old*) during the study phase by a button press and also to identify the images that had never been present (*new*) during the study phase by pressing another button. The reaction times were obtained with a three-button device placed in front of the subject. They were instructed to emphasize both speed and accuracy in their responses. The emotional evaluation based on the Self Assessment Manikin (SAM) was administered after the ERP experimentation and the participants rated, by a paper and pencil response, each of the 150 images presented in a booklet. Previous brain imaging studies using emotional photographic images have shown that task instructions, prompting preparation for the processing of the evocative images, are susceptible to affect neural activity. Thus, for both study and test phases, participants were not informed about the emotional value of the images beforehand in order to minimize emotional expectancy. The emotional evaluation of images was done post-test in order to keep the emotional nature of the task implicit during the experimentation.

4.5. EEG recordings and ERP extraction

The EEG was recorded from 28 tin electrodes mounted in an elastic nylon cap (Electro-Cap International Inc) only during recall (test phase). The scalp electrodes were placed according to the guidelines for standard electrode position by the at F3, FT7, FC3, T3, C3, F4, FC4, FT8, C4, T4, TP7, CP3, T5, P3, O1, CP4, TP8, P4, T6, O2. All electrodes were referenced to linked mastoids and their impedances were kept below 5 KΩ. The Electro-oculograms (EOG) was recorded using four 9-mm tin external bi-polar electrodes for horizontal and vertical movements. For the horizontal EOG, electrodes were placed at the outer canthus of each eye and for the vertical EOG at infra- and supra-orbital points at the left eye, aligned with the pupil looking straight. A bioelectric analog amplifier model ISS3-32BA (SAI-InstEP) amplified electrical signals (EOG gain = ±10000 and EEG gain = ±20000) with a band-pass between .01 and 30 Hz. The EEG was recorded continuously at a sampling rate of 250 Hz and averaged offline in a time-window beginning at 100 ms before and until 1900 ms after picture onset. The EOG artifact contained in the EEG were corrected with a dynamic multiple regression in the frequency domain. The regressions were applied using the horizontal and the vertical EOG activity subsequently. After EOG corrections, all remaining epochs with a voltage exceeding ±100 uV and clippings due to saturation or blocking of the amplifiers were eliminated automatically during the averaging procedure. On average, 2.5 trials per condition were rejected, after EOG corrections, because of the remaining artifacts (range = 0-5 trials). An analysis of variance (ANOVA), applied to the number of artefact rejected, failed to show any significant effect across response type and emotional valence conditions (all p's over .30). A second ANOVA applied on the two EOGs separately also failed to reach any statistical significance according to response type or valence (all p's > .10). A minimum amount of 16 trials free of both errors (false alarms and misses) and artifacts were included in the ERP averaging, which is comparable to the criteria used in similar ERP experiments. All ERP data were extracted for two time-windows (300-500ms and 500-1000ms). These two windows allowed us to study the EM effect, which usually appear between 300 and 800 ms after the stimulus presentation. The EM effect was first depicted by as a representation of recollection processes associated with medial temporal lobe structures. Our experimental hypotheses were tested using the mean amplitudes of the ERP detected within the temporal windows as defined in previous recognition memory research.

4.6. Statistical analysis

Several ANOVAs were performed on age, education and non-verbal intelligence (Raven matrices), as well as for BDI, BAI, Padua and Y-BOCS scores. Gender was analyzed using the Kruskall-Wallis, non-parametric test.

EM effect amplitude was analyzed separately using multivariate repeated-measures analyses of variance (MANOVA-RM). Subsequently, a separate multivariate analysis of covariance (MANCOVA) was applied on ERP amplitude data considering BDI and BAI as covariates. The analysis comprised a between-groups factor including three levels (OCD+, OCD- and control groups), and the following within-groups factors: Memory, with two levels (old, new); Emotion, with three levels (positive, neutral, negative); Hemisphere, with two levels (left, right); Region, with two levels (anterior, posterior) and Electrodes, with the five remaining levels. The electrodes were divided in four quadrants as left anterior (F3, FT7, FC3, T3 and C3), left posterior (TP7, CP3, T5, P3 and O1), right anterior (F4, FC4, FT8, C4 and T4) and right posterior (CP4, TP8, P4, T6 and O2).

5. Results

5.1. Subjective evaluation of emotional images

Activation and valence evaluation: On one hand, a non-significant memory factor indicated that memory showed no impact on the subjective evaluation of activation. On the other hand, the emotional valence influenced the subjective evaluation of activation [$F(2,38) = 151.17, p < 0.001$]. Thus, the subjective evaluation of activation was higher for positive (6.79/9) and negative (5.61/9) emotional valence stimuli than for neutral ones (3.82/9). Furthermore, we found a significant memory by emotion interaction [$F(2,38) = 25.50$, $p < 0.001$], indicating that the emotional valence effect was more important for the previously seen images. However, there was no group effect since participants from both OCD groups (OCD+ and OCD-) gave a subjective evaluation of activation that was equivalent to the control participants.

Dominance evaluation: The level of subjective confidence in participants' respective responses was also affected by the emotional valence of stimuli [$F(2,38) = 7.24$, $p < 0.005$]. We effectively observed that the dominance level was significantly higher for responses to positive images (7.93/9) compared to negative (7.53/9) and neutral ones (7.51/9). Additionally, a significant interaction between memory and emotions [$F(2,38) = 7.55$, $p < 0.005$] indicated that the subjective dominance level was greater only for responses to old positive images. Regarding the comparison group, there was a trend suggesting a lower subjective confidence level for both OCD groups when compared to the control group [$F(2,39) = 2.96$, $p = 0.06$].

5.2. Performance and reaction times

Reaction times: New images elicited a delayed reaction time (1329 ms) in comparison to old ones (1205 ms) [*Memory*: $F(1,42)$ 26.45, $p < 0.001$]. Likewise, reaction times to negative images (1307 ms) were significantly delayed when compared to positive (1249 ms) and neutral (1247 ms) ones [*Emotion*: $F(2,41) = 16.65$: $p < 0.001$]. As exposed before, we noticed differences between response times to old and new images. Thus, a significant memory by emotion interaction [$F(2,41) = 6.72$: $p < 0.005$] showed that these old-new differences were larger when a response was required for stimuli of negative emotional valence (158 ms) compare to positive (137 ms) and neutral (75 ms) ones (see Figure 1). Additionally, we have noticed that both OCD groups showed delayed reaction times than controls to the negative condition [*Group by emotion:* $F(2,42) = 3.84$, $p < 0.05$].

Analysis of correct responses (hits): The novelty aspect affected the amount of correct responses as indicated by a significantly greater amount of hits for new images than for old ones [*Memory:* $F(1,42) = 19.67$, $p < 0.001$]. As well, emotions also showed an impact on performances

[*Emotion:* F(1,42) = 19.67, p < 0.001], since negative images (24) elicited more hits than the neutral (23) and the positive (22) ones. Similarly, the emotional valence of images has also showed an impact on memory performances since a memory by emotion interaction was significant [F(2,41) = 27.99, p < 0.001] only in positive and neutral conditions. As for the reaction times analyses, a significant group by emotion interaction [F(4,84) = 3.34, p < 0.05] was found. This indicates that negative images evoked better performances for the control and OCD- groups, as represented by a higher number of hits.

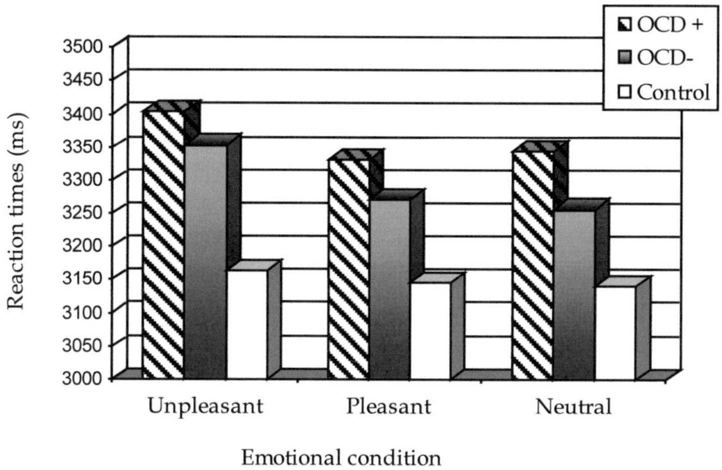

Figure 1. Illustration of the group by emotion interaction on the reaction times (RT) in function of the three emotional conditions for the three groups. The RTs were delayed more dramatically in the unpleasant condition for both OCD groups compared to the control group.

5.3. Electrophysiological data

300 ms to 500 ms post-stimulus window: The topographic distribution of ERPs revealed greater amplitudes in the posterior comparatively to the anterior area [F (1,42) = 49.28: p < 0.001]. As well, emotional valence impacted electrophysiological data as shown by more salient ERPs in response to positive and negative images than those for the neutral ones [Emotion: F (2,41) = 16.95: p < 0.001]. Similarly, memory also influenced 300-500 ms brain activity as suggested by significantly larger ERPs for old images than for new ones [F (1,42) = 30.34: p < 0.001], thus attesting for the early EM effect. Additionally, a complex interaction between memory, emotion and regions reached significance [F (2,41) = 3.57: p < 0.05]. Accordingly, specifically over the anterior region, the EM effect was more important to positive and neutral emotions than to negative ones. Furthermore, no effect of memory or emotions was evident over the posterior region and importantly, no group difference was found for this time-window.

500 ms to 1000 ms post-stimulus window: The emotion factor remained significant in our second time-window [F (2,41) = 16.31: p < 0.001]. The ERPs to positive and negative images were significantly larger compared to those in response to neutral ones. Additionally, a complex interaction between memory, emotion and hemispheres was present [F (2,41) = 4.14: p < 0.05]. When taking into account the memory factor, ERPs in response to neutral images were larger in the left hemisphere comparatively to the right hemisphere. Hemispheric lateralization of the memory effect as observed in the left hemisphere was also noted for the negative images (without being as important as for neutral and positive images). Similarly to the first time-window, the ERPs scalp distribution revealed larger amplitudes in the posterior region comparatively to the anterior area [F (1,42) = 13.38: p = 0.001]. A memory by region interaction [F (1,40) = 7.87: p < 0.01] indicated that the amplitudes differences between old and new images were greater in the posterior region than in the anterior area. Additionally, OCD+ participants showed reduced amplitudes in comparison to controls and OCD- groups particularly in the posterior region, as suggested by a significant region and group interaction [F (2,42) = 5.49: p < 0.010]. Moreover, the group by memory interaction remained significant after covarying for anxiety scores [F (2,40) = 3.78: p < 0, 05].

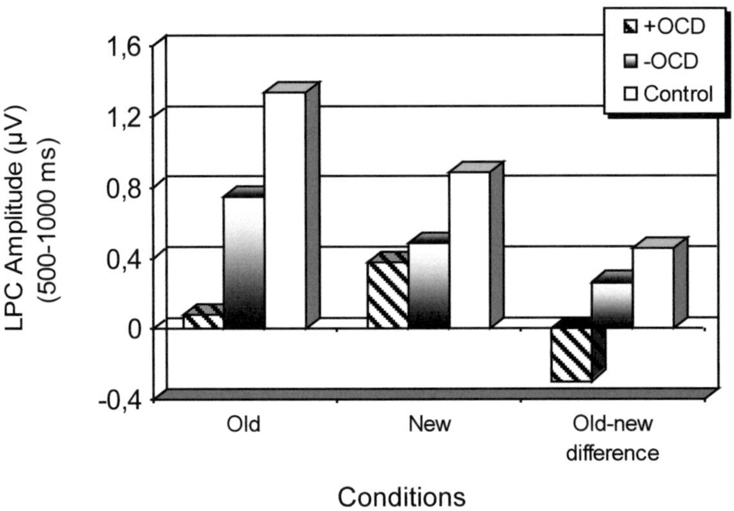

Figure 2. Illustration of the group by memory interaction with the LPC amplitude in function of old-new memory condition for the three groups.

The EM effect reflected by the difference between old and new stimuli was significantly smaller for OCD- and OCD+ respectively than for the control group, as shown in Figure 2. Also, a four-way complex interaction between groups, memory, emotions and regions has been noticed [F(2,40) = 3.57, p < 0.05].

6. Discussion

Globally, analyses suggested that old pictures triggered faster reaction times and that OCD groups were characterized by delayed reaction times and less correct responses comparatively

http://dx.doi.org/10.5772/57179

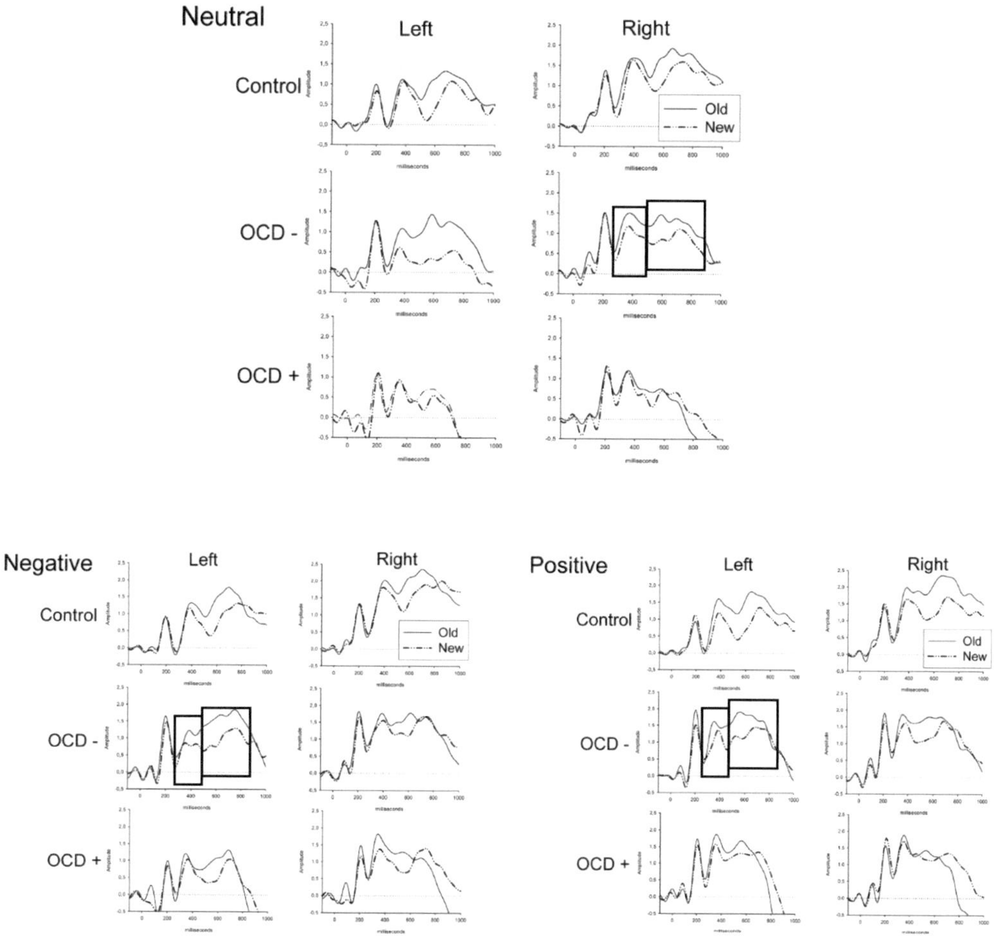

Figure 3. Illustration of the stimulus-locked ERP amplitudes to old (solid lines) and new (dotted lines) images for all groups in neutral, negative and positive valence conditions. The left column represent ERPs of the left posterior parietal region and the right column refer to ERPs of the right posterior parietal region. The dotted squares represent the two time windows of interest (300-500 ms and 500-1000 ms post-stimulus). The OCD+ showed a reduced old-new effect compared to the OCD- and to the controls.

to healthy participants. These observations are consistent with current and past research investigating anxious and depressive patients who showed similar phenomena. Additionally, the pictures' emotional valence influenced all participants' episodic memory processing. As well, the EM effect was smaller for all OCD patients, and even more particularly for the most severely affected group. The severity of OCD symptoms seems to affect brain activity related to memory processing and this could be, at least in part, explained by a dysfunction in the CSTC. Moreover, emotional pictures elicited a reduced anterior EM effect among severe OCD patients. Additionally, the insertion of depressive or anxious symptoms as covariables did not influence the results, suggesting that the presence of significant comorbid symptoms were not related to the altered emotional memory effect observed in OCD patients.

6.1. Emotional memory performances and behavior

In general, performances and response times were delayed and the number of correct responses was higher for new images. Similarly, it has been shown that new words elicit delayed reaction times than old ones. Likewise, in a study investigating episodic memory accounting for emotional factors of Gilles de la Tourette syndrome and OCD (SGT-OCD) patients, also found that old words, that were presented only once, elicited shorter reaction times. Conversely, found that depressive patients presented delayed reaction times for old words. Unlike, we found that the emotional valence did in fact, influenced reaction times. Indeed, our results showed that overall, negative images were those that elicited the fastest reaction times. Additionally, participants were faster to detect old negative and positive images comparatively to new positive and negative images. Using words rather than photographs, also reported the same findings in depressive patients.

Analogously, the amount of correct responses was also influenced by the emotional valence of images, especially for newly presented neutral and positive ones. This suggests that emotional valence influences performance and modulates memory processes. Additionally, we observed a trend toward longer reaction times for OCD patients when compared to controls. Consistently, observed the same phenomenon in different OCD patients. Likewise, when they compared moderately symptomatic OCD patients to controls, they showed significantly delayed reaction times. These results coincide with those found by during an implicit memory task that was part of an experiment assessing OCD patients' memory using words. It is interesting to note that the OCD patients included in that study were comparable to our less symptomatic patients (OCD-) and that similar results were found (i.e. patients and controls showed delayed reaction times for old words. Moreover, we also found that, for images of the same emotional category, OCD+ participants showed a delayed response when compared to controls. Additionally, our more symptomatic OCD patients tended to be less confident in their responses to negative images compared to controls. These group difference could be explained by a different recall strategy aiming to compensate the difficulties experienced by more severe OCD symptoms. From past results, we also know that checker OCD patients have a poorer perception of their memory capacity and that they lack confidence in their recognition capabilities.

6.2. The influence of OCD symptoms on episodic memory and ERPs

Between 300 and 500 ms post-stimulus, old images elicited larger ERP amplitude than new images. Our results strongly resemble those of, which demonstrated that ERPs were more positive for repeated words (i.e. old words). This old/new effect has been shown to be reflected by episodic memory and is sensitive to conscious recollection. In our second time-window (500-1000 ms post-stimulus), we also found that OCD+ participants showed significantly smaller ERPs for old images. Moreover, this result remained significant even after covarying for anxiety level.

Conversely, Kim et al. (2006) found no difference between their OCD patients and controls for the EM effect with words. However, their OCD patients were moderately symptomatic (Y-

BOCS = 25), and resembled more to our OCD- patients (Y-BOCS = 24) than our OCD+ patients (Y-BOCS = 34+). In fact, our OCD- group revealed no significant differences compared to the control group regarding the magnitude of the EM effect. These results suggest that only highly symptomatic OCD patients may recollect information differently. Indeed, it has been found that people suffering from OCD may feel the need to adopt a sequential rather than a comprehensive approach to recognition and organization when performing even a simple memory task. A possible explanation regarding the results discrepancies between different studies may lie in the severity of OCD symptoms.

Another interesting finding is that ERPs remained more important in the posterior region compared to the anterior area. This could indicate the presence of a late positive component (LPC) associated with episodic memory. The EM effect is associated with the conscious recollection processes that occur primarily in the left parietal scalp region, normally between 400 and 800 ms. Interestingly, we observed, comparatively to OCD- and controls, that highly symptomatic OCD+ patients showed smaller EM effects, particularly in the posterior region. Likewise, performed a similar study, comparing severe and moderate OCD patients to control participants. They studied the EM effect after presenting new and old words, while recording ERP signals. Although our OCD+ group (mean Y-BOCS = 33) was more symptomatic than their severe group (mean Y-BOCS = 27), they also observed a diminution of the late EM effect (450-650 ms). This reduction was specific to the severe OCD group compared to moderate OCD and controls and was also more prominent in the parieto-temporal region. This could indicate that severe OCD symptoms impact on processes mediated through the right prefrontal cortical regions, which are hypothesized to be involved in memory inhibition mechanisms.

6.3. The influence of OCD on episodic memory and the influence of emotions on ERPs

With our participants, we noted that positive and negative images evoked the largest ERPs. As well, numerous researchers have also reported that they observed increased ERPs' amplitudes when presenting stimuli with an emotional component. Moreover, added that emotions might modulate recognition processes. Thus, it is possible that ERPs reflect, in part, the activity of the anterior cingulate cortex, which has connections with cortico-frontal regions and limbic system. Additionally, found a decrease of the EM effect when they presented words with positive and negative connotations to SGT-OCD patients. In parallel, for our first-time window (300-500 ms post-stimulus), neutral images triggered larger EM effects and positive ones elicited even larger ERPs. Analogously, reported similar conclusions following their SGT-OCD study, where ERPs were larger particularly in the frontal and central regions.

Regarding our second time-window, the memory effect was linked to larger ERPs in the left hemisphere particularly for neutral and negative images. This result seems consistent with other findings with verbal stimuli. Recognition information that has no emotional component might be related to more limited cerebral regions while information with emotional factors might be related to other structures. found a smaller EM effect for positive images for SGT-OCD patients comparatively to control participants. While covarying for anxiety levels, we have observed the same phenomenon more particularly over the posterior region for OCD+

and controls. Our results lead us to the same hypothesis articulated by: patients showed diminished capacities for conscious recollection than healthy controls, especially in an emotionally charged context.

7. Conclusion

Our findings highlighted the influence of severe OCD symptoms on processes mediated through the right prefrontal cortical regions, which are hypothesized to be involved in memory inhibition mechanisms. Indeed, it has been found that people suffering from OCD may feel the need to adopt a sequential rather than a comprehensive approach to recognition and organization when performing even a simple memory task. Anomaly in certainty or memory inhibition mechanisms could influence information processing and working memory updating processes, as reflected by an anterior P300 amplitude attenuation. Thus, our findings support the importance of selecting more symptomatic participants in order to study differences in emotional memory processes. This might explain why a relatively important part of the literature often failed to notice any significant difference between OCD and control participants.

Acknowledgements

This work was supported in part by a Canadian Institutes of Health Research (CIHR) operating grant (MOP57936), and a Fonds de Recherche du Québec - Santé (FRQS) team research grant *(Subvention à la recherche en santé mentale -FRQS #20573).* Geneviève Sauvé was supported by a Graduate student recruitment scholarship from the Faculty of Medicine, University of Montreal and a Master scholarship from the IUSMM foundation. Simon Morand-Beaulieu was supported by a Master scholarship from the biomedical sciences program of the University of Montreal.

Author details

Marc E. Lavoie, Geneviève Sauvé, Simon Morand-Beaulieu, Marie-Pierre Charron and Kieron P. O'Connor

Cognitive and Social Psychophysiology Laboratory, Multidisciplinary team on OCD spectrum, Centre de Recherche de l'Institut Universitaire en Santé Mentale de Montréal, Department of Psychiatry, University of Montréal, Québec, Canada

References

[1] American EEG Society. (1994). Guideline thirteen: guidelines for standard electrode position nomenclature. American Electroencephalographic Society. *J Clin Neurophysiol, 11*(1), 111-113.

[2] Andres, S., Boget, T., Lazaro, L., Penades, R., Morer, A., Salamero, M., & Castro-Fornieles, J. (2007). Neuropsychological performance in children and adolescents with obsessive-compulsive disorder and influence of clinical variables. *Biol Psychiatry, 61*(8), 946-951. doi: 10.1016/j.biopsych.2006.07.027

[3] Beck, A. T., Epstein, N., Brown, G., & Steer, R. A. (1988). An inventory for measuring clinical anxiety: psychometric properties. *Journal of Consulting and Clinical Psychology, 56*(6), 893-897.

[4] Beck, A. T., Ward, C. H., Mendelson, M., Mock, J., & Erbaugh, J. (1961). An inventory for measuring depression. *Archives of General Psychiatry, 4*, 561-571.

[5] Beech, H. R., Ciesielski, K. T., & Gordon, P. K. (1983). Further observations of evoked potentials in obsessional patients. *Br J Psychiatry, 142*, 605-609.

[6] Bohne, A., Savage, C. R., Deckersbach, T., Keuthen, N. J., Jenike, M. A., Tuschen-Caffier, B., & Wilhelm, S. (2005). Visuospatial abilities, memory, and executive functioning in trichotillomania and obsessive-compulsive disorder. *J Clin Exp Neuropsychol, 27*(4), 385-399. doi: 10.1080/13803390490520418

[7] Boldrini, M., Del Pace, L., Placidi, G. P., Keilp, J., Ellis, S. P., Signori, S.,... Cappa, S. F. (2005). Selective cognitive deficits in obsessive-compulsive disorder compared to panic disorder with agoraphobia. *Acta Psychiatr Scand, 111*(2), 150-158. doi: 10.1111/j.1600-0447.2004.00247.x

[8] Bradley, M. M., & Lang, P. J. (1994). Measuring emotion: the Self-Assessment Manikin and the Semantic Differential. *J Behav Ther Exp Psychiatry, 25*(1), 49-59.

[9] Burdick, K. E., Robinson, D. G., Malhotra, A. K., & Szeszko, P. R. (2008). Neurocognitive profile analysis in obsessive-compulsive disorder. *J Int Neuropsychol Soc, 14*(4), 640-645. doi: 10.1017/S1355617708080727

[10] Constans, J.I, Foa, E.B., Franklin, M.E., & Mathews, A.. (1995). Memory for actual and imagined events in OC checkers. *Behavioral Research and Therapy, 33*(6), 665-671.

[11] Cougle, J. R., Salkovskis, P. M., & Wahl, K. (2007). Perception of memory ability and confidence in recollections in obsessive-compulsive checking. *J Anxiety Disord, 21*(1), 118-130. doi: 10.1016/j.janxdis.2006.03.015

[12] Curran, T., & Friedman, W. J. (2003). Differentiating location- and distance-based processes in memory for time: an ERP study. *Psychon Bull Rev, 10*(3), 711-717.

[13] de Geus, F., Denys, D. A., Sitskoorn, M. M., & Westenberg, H. G. (2007). Attention and cognition in patients with obsessive-compulsive disorder. *Psychiatry Clin Neurosci, 61*(1), 45-53. doi: 10.1111/j.1440-1819.2007.01609.x

[14] Deckersbach, T., Savage, C. R., Reilly-Harrington, N., Clark, L., Sachs, G., & Rauch, S. L. (2004). Episodic memory impairment in bipolar disorder and obsessive-compulsive disorder: the role of memory strategies. *Bipolar Disord, 6*(3), 233-244. doi: 10.1111/j.1399-5618.2004.00118.x

[15] Delis, D.C., Kramer, J.H., Kaplan, E., & Ober, B.A. (2000). *California Verbal Learning Test: Second Edition*. San Antonio, TX: Psychological Corporation.

[16] Depue, B. E., Curran, T., & Banich, M. T. (2007). Prefrontal regions orchestrate suppression of emotional memories via a two-phase process. *Science, 317*(5835), 215-219. doi: 10.1126/science.1139560

[17] Dietrich, D. E., Emrich, H. M., Waller, C., Wieringa, B. M., Johannes, S., & Munte, T. F. (2000). Emotion/cognition-coupling in word recognition memory of depressive patients: an event-related potential study. *Psychiatry Res, 96*(1), 15-29.

[18] Dolcos, F., & Cabeza, R. (2002). Event-related potentials of emotional memory: encoding pleasant, unpleasant, and neutral pictures. *Cogn Affect Behav Neurosci, 2*(3), 252-263.

[19] Friedman, D., & Johnson, R., Jr. (2000). Event-related potential (ERP) studies of memory encoding and retrieval: a selective review. *Microsc Res Tech, 51*(1), 6-28. doi: 10.1002/1097-0029(20001001)51:1<6::AID-JEMT2>3.0.CO;2-R

[20] Goodman, W. K., Price, L. H., Rasmussen, S. A., Mazure, C., Delgado, P., Heninger, G. R., & Charney, D. S. (1989). The Yale-Brown Obsessive Compulsive Scale. II. Validity. *Arch Gen Psychiatry, 46*(11), 1012-1016.

[21] Graybiel, A. M., & Rauch, S. L. (2000). Toward a neurobiology of obsessive-compulsive disorder. *Neuron, 28*(2), 343-347.

[22] Hamann, S. B., Ely, T. D., Grafton, S. T., & Kilts, C. D. (1999). Amygdala activity related to enhanced memory for pleasant and aversive stimuli. *Nat Neurosci, 2*(3), 289-293. doi: 10.1038/6404

[23] Johannes, S. (1999). Evidence for changed recognition of emotionally charged words in patients with Gilles de la Tourette syndrome and obsessive compulsive disorder. *Cogn Neuropsychiatry, 4*(1), 37-53. doi: 10.1080/135468099396052

[24] Johnson, R., Jr., Kreiter, K., Russo, B., & Zhu, J. (1998). A spatio-temporal analysis of recognition-related event-related brain potentials. *Int J Psychophysiol, 29*(1), 83-104.

[25] Kim, M. S., Kim, Y. Y., Kim, E. N., Lee, K. J., Ha, T. H., & Kwon, J. S. (2006). Implicit and explicit memory in patients with obsessive-compulsive disorder: an event-related potential study. *J Psychiatr Res, 40*(6), 541-549. doi: 10.1016/j.jpsychires.2005.05.007

[26] Lang, P.J., Bradley, M.M., & Cuthbert, B.N. (1997). *International Affective Picture System (IAPS): Technical Manual and Affective Ratings.* Gainesville, FL: NIMH Center for the study of Emotion and Attention, University of Florida.

[27] Lavy, E., van Oppen, P., & van den Hout, M. (1994). Selective processing of emotional information in obsessive compulsive disorder. *Behav Res Ther, 32*(2), 243-246.

[28] Lee, A. K., Larson, E., Maddox, R. K., & Shinn-Cunningham, B. G. (2013). Using neuroimaging to understand the cortical mechanisms of auditory selective attention. *Hear Res*. doi: 10.1016/j.heares.2013.06.010

[29] Maratos, E. J., Allan, K., & Rugg, M. D. (2000). Recognition memory for emotionally negative and neutral words: an ERP study. *Neuropsychologia, 38*(11), 1452-1465.

[30] Mataix-Cols, D., Rahman, Q., Spiller, M., Alonso, M. P., Pifarre, J., Menchon, J. M., & Vallejo, J. (2006). Are there sex differences in neuropsychological functions among patients with obsessive-compulsive disorder? *Appl Neuropsychol, 13*(1), 42-50. doi: 10.1207/s15324826an1301_6

[31] Morault, P. M., Bourgeois, M., Laville, J., Bensch, C., & Paty, J. (1997). Psychophysiological and clinical value of event-related potentials in obsessive-compulsive disorder. *Biol Psychiatry, 42*(1), 46-56. doi: 10.1016/S0006-3223(96)00228-4

[32] Moritz, S., Kloss, M., Jahn, H., Schick, M., & Hand, I. (2003). Impact of comorbid depressive symptoms on nonverbal memory and visuospatial performance in obsessive-compulsive disorder. *Cogn Neuropsychiatry, 8*(4), 261-272. doi: 10.1080/135468000344000020

[33] Oltmanns, T. F., & Gibbs, N. A. (1995). Emotional responsiveness and obsessive-compulsive behaviour. *Cognition & Emotion, 9*(6), 563-578. doi: 10.1080/02699939508408983

[34] Palomba, D., Angrilli, A., & Mini, A. (1997). Visual evoked potentials, heart rate responses and memory to emotional pictorial stimuli. *Int J Psychophysiol, 27*(1), 55-67.

[35] Papageorgiou, C. C., & Rabavilas, A. D. (2003). Abnormal P600 in obsessive-compulsive disorder. A comparison with healthy controls. *Psychiatry Res, 119*(1-2), 133-143.

[36] Penades, R., Catalan, R., Andres, S., Salamero, M., & Gasto, C. (2005). Executive function and nonverbal memory in obsessive-compulsive disorder. *Psychiatry Res, 133*(1), 81-90. doi: 10.1016/j.psychres.2004.09.005

[37] Polich, J. (2007). Updating P300: an integrative theory of P3a and P3b. *Clin Neurophysiol, 118*(10), 2128-2148. doi: 10.1016/j.clinph.2007.04.019

[38] Radomsky, A. S., & Rachman, S. (1999). Memory bias in obsessive-compulsive disorder (OCD). *Behav Res Ther, 37*(7), 605-618.

[39] Radomsky, A. S., Rachman, S., & Hammond, D. (2001). Memory bias, confidence and responsibility in compulsive checking. *Behav Res Ther, 39*(7), 813-822.

[40] Rugg, M. D., & Nagy, M. E. (1989). Event-related potentials and recognition memory for words. *Electroencephalogr Clin Neurophysiol, 72*(5), 395-406.

[41] Sadia, G., Ritter, W., & Sussman, E. (2013). Category effects: is top-down control alone sufficient to elicit the mismatch negativity (MMN) component? *Biol Psychol, 92*(2), 191-198. doi: 10.1016/j.biopsycho.2012.10.008

[42] Sanavio, E. (1988). Obsessions and compulsions: the Padua Inventory. *Behaviour research and therapy, 26*(2), 169-177.

[43] Savage, C. R., Baer, L., Keuthen, N. J., Brown, H. D., Rauch, S. L., & Jenike, M. A. (1999). Organizational strategies mediate nonverbal memory impairment in obsessive-compulsive disorder. *Biol Psychiatry, 45*(7), 905-916.

[44] Savage, C. R., & Rauch, S. L. (2000). Cognitive deficits in obsessive-compulsive disorder. *Am J Psychiatry, 157*(7), 1182-1183.

[45] Simpson, H. B., Rosen, W., Huppert, J. D., Lin, S. H., Foa, E. B., & Liebowitz, M. R. (2006). Are there reliable neuropsychological deficits in obsessive-compulsive disorder? *J Psychiatr Res, 40*(3), 247-257. doi: 10.1016/j.jpsychires.2005.04.004

[46] Steketee, G. (1994). Behavioural assessment and treatment planning with obsessive-compulsive disorder. *Behavioral Therapy, 25*, 613-633.

[47] Tabert, M. H., Borod, J. C., Tang, C. Y., Lange, G., Wei, T. C., Johnson, R.,... Buchsbaum, M. S. (2001). Differential amygdala activation during emotional decision and recognition memory tasks using unpleasant words: an fMRI study. *Neuropsychologia, 39*(6), 556-573.

[48] Taylor, S. (1995). Assessment of obsessions and compulsions: reliability, validity, and sensitivity to treatment effects. *Clinical Psychology Review., 15*, 261-297.

[49] Taylor, S. F., Phan, K. L., Decker, L. R., & Liberzon, I. (2003). Subjective rating of emotionally salient stimuli modulates neural activity. *Neuroimage, 18*(3), 650-659.

[50] Thibault, G., Felezeu, M., O'Connor, K. P., Todorov, C., Stip, E., & Lavoie, M. E. (2008). Influence of comorbid obsessive-compulsive symptoms on brain event-related potentials in Gilles de la Tourette syndrome. *Prog Neuropsychopharmacol Biol Psychiatry, 32*(3), 803-815. doi: 10.1016/j.pnpbp.2007.12.016

[51] Towey, J. P., Bruder, G., Hollander, E., Friedman, D., Erhan, H., Liebowitz, M., & Sutton, S. (1990). Endogenous event-related potentials in obsessive-compulsive disorder. *Biol Psychiatry, 28*(2), 92-98.

[52] Towey, J. P., Bruder, G., Tenke, C., Leite, P., DeCaria, C., Friedman, D., & Hollander, E. (1993). Event-related potential and clinical correlates of neurodysfunction in obsessive-compulsive disorder. *Psychiatry Res, 49*(2), 167-181.

[53] Towey, J. P., Tenke, C. E., Bruder, G. E., Leite, P., Friedman, D., Liebowitz, M., & Hollander, E. (1994). Brain event-related potential correlates of overfocused attention in obsessive-compulsive disorder. *Psychophysiology, 31*(6), 535-543.

[54] Unoki, K., Kasuga, T., Matsushima, E., & Ohta, K. (1999). Attentional processing of emotional information in obsessive-compulsive disorder. *Psychiatry Clin Neurosci, 53*(6), 635-642. doi: 10.1046/j.1440-1819.1999.00618.x

[55] Wilhelm, S., McNally, R. J., Baer, L., & Florin, I. (1996). Directed forgetting in obsessive-compulsive disorder. *Behav Res Ther, 34*(8), 633-641.

[56] Windmann, S., & Kutas, M. (2001). Electrophysiological correlates of emotion-induced recognition bias. *J Cogn Neurosci, 13*(5), 577-592. doi: 10.1162/089892901750363172

[57] Woestenburg, J. C., Verbaten, M. N., & Slangen, J. L. (1983). The removal of the eye-movement artifact from the EEG by regression analysis in the frequency domain. *Biol Psychol, 16*(1-2), 127-147.

[58] Zhang, Y., Feutl, S., Hauser, U., Richter-Witte, C., Schmorl, P., Emrich, H. M., & Dietrich, D. E. (2008). Clinical correlates of word recognition memory in obsessive-compulsive disorder: an event-related potential study. *Psychiatry Res, 162*(3), 262-272. doi: 10.1016/j.pscychresns.2007.04.009

Section 2

Therapeutic Approaches to the Treatment of Obsessive-Compulsive Disorder

Chapter 5

Cognitive-Behavior and Narrative Therapy in Obsessive-Compulsive Disorder

Mª-José Martín-Vázquez

Additional information is available at the end of the chapter

http://dx.doi.org/10.5772/57579

1. Introduction

From its very beginning, behavior-cognitive therapy has been applied in the treatment of obsessive-compulsive disorder (OCD), particularly using the exposure with response prevention strategy.

This treatment has different components: one of them is the exposure to the situation that usually causes the compulsive behavior without completing the compulsive acts. The other is cognitive exposure to the secondary anxiety caused by the avoiding compulsory activity, with the confirmation that the feared situation doesn't come.

Cognitive Behavior Therapy (CBT) is as efficient as pharmacotherapy in the treatment of OCD and the effects last at least the same time. Some studies have found that the efficacy of the combination of CBT and medication is higher than that of each one separately. But the studies vary regarding the number of patients who achieved remission. Rodrigues et al [2011] revised published studies with CBT and pharmacotherapy and they found than the percentages of remission varied from 64% to 23%. But all the studies revised in their systematic review suggested the efficacy of CBT as a next-step strategy for treating patients who do not remit with pharmacological therapy only [1].

Almost 60% of patients with OCD do not respond adequately to medications and are considered to be refractory to pharmacotherapy [2]. Relapse rates are still high among OCD patients undergoing pharmacotherapy, from 24% to 89%, in part because of differences in study designs and definitions of relapse. CBT provides promise for OCD patients, with effectiveness rates ranging from 60% to 85%. Relapse rates three months after discontinuation of intensive CBT are up to 50%.

Desinsitization, hought detention, flooding, implosion therapy, and aversive conditioning have been used in the treatment of OCD; all of them are based on the collaboration of the patient and the need of tolerating anxiety related to obsessive thoughts, avoiding acting the compulsions.

Following psychoanalytic view, OCD is related to aggressive impulses; so accepting them through introspective or cognitive therapy can produce positive results. Constructive psychotherapy can help some patients with good cognitive capacity. Cognitive component of therapy probably works emphasizing the creation of new representations and behaviors more than correcting distorted cognitions.

In some cases group therapy is useful; it has been proved in children and adolescent patients, populations that usually have problems to accept they are different to their peers. Group therapy enhances treatment adherence and facilitates the use of humor in therapy, particularly relevant for OCD patients because they live in a fearful and pessimistic world, and reveal a significant lower level of positive emotions, even compared with depressive population.

During cognitive therapy obsessions and compulsions are defined and relabeled as OCD symptoms, different from ordinary thoughts. Patients are encouraged to distance themselves from these obsessions and view these thoughts as strange to them. The exposure has to be planned, with a previous exposure hierarchy. Refocusing is a way of increasing tolerance for response prevention; refocusing on an alternative behavior facilitates response prevention and thought detention.

CBT procedures involve the selection and creation of alternative representations to restore more positive mood states. New directions in CBT try to modify a person's relationship to his/her negative thoughts rather than directly to challenge the content of the thoughts.

Mindfulness-based cognitive therapy is based on the application of Buddhist meditation techniques and it is beginning to be used in OCD as a complementary technique in treatment.

2. Obsessive-Compulsive Disorder — A brief explanation from cognitive approach

Obsessive-compulsive disorder (OCD) is considered a multifactorial disorder that can significantly impair functionality and quality of life [3]. Though obsessive thoughts are common in general population, in OCD the amount of time dedicated to them impairs the execution of normal functioning.

Patients with OCD selectively attend to fearful signs, but they generally present difficulties to inhibit irrelevant information as well. OCD is characterized for the preconscious deficit in the ability to process irrelevant information which is essential to symptoms formation and maintenance [4].

http://dx.doi.org/10.5772/57579

Compulsions can involve washing, ordering, cleaning, checking, seeking reassurance, repetitive actions as rituals, or mental compulsions, and their objective uses to be preventing some danger to oneself or to the loved ones, or avoiding anxiety provoked by the obsessions. Obsessions are found to be differentiated into five factors: contamination/cleaning, symmetry/ordering, aggressive/checking, pure obsessions, and hoarding [5]. Checking repetitions could be an altered function of memory of emotional events, a disability to distinguish what is the imaginary part of events or a deficient record without alterations of recent memory [4]. Repetitive checking behaviors are considered an example of the lack of certainty over the realization or not of an action. The patients cannot distinguish between "have done it" or "have imagined doing it". This uncertainty could lead to obsessive doubts that provoked repetitive checking of doors, gas key, or others [4].

Classically OCD patients are classified as "washers" or "checkers", but classified with five factors seems to be more explanatory. Cognitive vulnerability dimensions may vary according to symptom subtype: harming-related fears due to the spread of contamination patients have higher scores in responsibility/thread estimation when compared with patients with pure contamination symptoms [5]. Inflated responsibility beliefs are related to harming/doubting/checking obsessive compulsive symptoms. Responsibility appraisals are less relevant for washers than for checkers or patients with aggressive/harming obsessions. Patients with contamination obsessive compulsive symptoms demonstrate cognitive bias to thematically related symptom stimuli, but patients with harming obsessions do not evidence greater cognitive bias to responsibility words than contamination patients [5]. Contamination patients have poorer response to cognitive therapy than checkers, which has to be kept in mind when a treatment is planned.

Memories do not preserve a literal representation of the world [6], and memory retrieval is constructive. In OCD it is frequent to observe illusory memories about previous danger that have been avoided thanks to rituals. Long-term memories are influenced by the emotion experienced during learning [7]. Mood-dependent memory is the enhance recollection of information previously recorded in a specific mood state. In patients with OCD, neutral information about cleanliness, sexuality, or order, is invested with a strong emotional correlate and becomes a fixed and obsessive idea. The internal representation of previously experienced emotional stimulus may elicit a transient emotional state [7]. This mechanism impairs the efficacy of cognitive therapy, because if the therapeutic work is to reconsolidate a new memory trace, an emotional stimulus is needed. Cognitive-behavior therapy acts on the premise that emotional memories are modified upon retrieval. In narrative approach, the retrieval is used, but the objective is not to reduce anxiety thanks to be exposed to the memory in a safe environment. The final objective is to re-explain the past experiences with a new look.

People with OCD tend to present deficit in tasks which imply cognitive bias and distortions, and they can be more sensitive in front of stimuli related to their fears [8], though contradictory data are found. Affective images are more capable to provoke cognitive bias than neutral ones. People with subclinical obsessive-compulsive symptoms evaluate neutral images less amiable than control people do, and all type of images as less controllable. OCD people need to feel control over every aspect of their lives.

Obsessions are caused by catastrophic misinterpretations of the significance of one's thoughts or impulses, similar to normal negative intrusions but more severe, frequent and upsetting [9]. There are two stages in the model: first an intrusion and then the misinterpretation of it. The OCD problems surge with the combination of perceived high responsibility and expectation of a future catastrophe. Some authors view responsibility as central in OCD, but others suppose it's secondary to other processes.

The feared scenario appears more easily in OCD imagination than non-feared scenarios. The repetition and "practice" of the same obsession results in the relevant simulation becoming more coherent and elaborated and reinforced through ritualizing, fixing the symptomatology [10].

Thinking errors that lead to obsessions can be derived from "fusions", as moral thought-action fusion (thinking of an action is as bad as doing it), thought-event fusion (thinking about an event provokes it), thought-thought fusion (thinking about having a thought is the same that having it) [9].

Criticism was defined as "a negative reinforcement which produces feelings of failure and has been identified as a poor way to encourage better performance" (Neapolitan, cited in 3). Among the difficulties in definition, criticism can mean different things for different people, from "an act of analysis" to having quite negative connotations. Self-criticism has been investigated in relation to mental health, and depends on feedback from others and internal thoughts.

Central to the cognitive model of OCD is the role of responsibility. The distress experienced in OCD has relation with the cognition of the self-responsibility about causing danger to oneself or the others. People with OCD need feeling control over external circumstances, managing them through rituals and compulsions.

Beliefs related to inflated responsibility have been proposed as one of six cognitive variables that play a role in OCD (Obsessive Compulsive Cognitions Working Group). The other five ones are: overimportance of thoughts (Thought Action Fusion), excessive concern about the importance of controlling one's thoughts, overestimation of threat, intolerance of uncertainty, and perfectionism [3].

The cognitive model of OCD emphasizes the role of early experiences that could predispose an individual to develop an OCD. Rituals and compulsive behaviors can appear as a way to avoid external criticism, and gain paternal approval. So, OCD could be the result of social training. The parenting style in families with OCD people uses to be overprotective and critical what could increase the need of control and responsibility. The experiences of recurrent criticism may increase the "subjective cost of being responsible". Harm eventually is described by OCD patients as excessive parental criticism. Overprotective parents may result in fearful children. If overcritical style is added, the result could be children who attempt to control and do everything right and avoid errors, resulting in checking behaviors. OCD parents' relationships frequently are characterized by perfectionism, high level of criticism and risk-aversion. It is frequent to observe familiar antecedents of compulsive behaviors or obsessive thoughts in OCD. Though it may be related to genetic factors, cognitive and behavioral styles are usually learned at home.

Perfectionism and responsibility functions are to maintain a sense of self-worth, maintaining approval from others, specifically to gain approval from hypercritical parents. Individuals with OCD may feel unsure about their self-worth due to receiving contradictory messages from a dominant parent during childhood. Dysfunctional responsibility and responsibility beliefs were related to an ambivalent sense of self. The cognitive model of OCD proposes that an individual fears harm coming to self or being responsible and blamed for causing harm. Compulsive behaviors may emerge as a way to regain a positive self-image or to retain social approval.

Responsibility in OCD uses to be high, and they are told frequently "you will be to blame if anything happens". When responsibility decreases, also anticipated criticism does.

Compulsive behaviors increase when an expectation of failure is present as well. There is a relationship between criticism and checking behaviours. These are preventive, though the cleaning ones are restorative. Through interactions with parents children learn to regulate themselves, so excessive criticism for mistakes or punishment for irresponsibility incites a concern with safety and responsibility.

3. Cognitive behavior therapy

CBT procedures involve the selection and creation of alternative representations to modify irrational ways of thinking and dysfunctional ways of behaving [11]. Desensitization and exposure have been largely used in the treatment of OCD patients. CBT has become the psychological treatment of choice for OCD, reducing symptoms to a level similar to that seen with pharmacotherapy, with clinical improvement maintained during follow-up [12]. In many studies CBT associated to medication management has demonstrated to be superior to the two strategies separately, even in pediatric age [13], though it seems to be more efficient the addition to CBT to people with a partial response to medication.

An obsession is both meaningful and irrational, part of the self and yet alien and intrusive [9]. For this reason, obsessions are a problem for cognitive approaches.

But CBT does not always lead to clinical improvement. Sometimes even secondary effects can appear during CBT, as nausea and abdominal discomfort, which are related to the anxiety provoked by the exposition and response prevention along the treatment [12]. New directions in cognitive approach attempt to change the persons' relationships with their thoughts, not to directly intervene over the thought content. The cognitive models usually assume that previous adversity provokes vulnerability in the form of negative representations of the self and the world (negative schemas). Selves can be based on wishes or aspirations, and in OCD, as in other anxiety disorders, symptoms are associated with a perceived failure to be the person oneself thinks he or she ought to be. Also, anxiety can be related to feeling too close to a feared or undesired self, and striving to avoid experiencing it [11], particularly relevant to OCD. Learning is a constructive process, constantly producing new representations which can collaborate or compete with the preexisting memories to control behavior, so the construction of new representations can be more efficient in the correction of distortions.

In OCD treatment can be focused in reduction of anxiety level, understanding OCD as an anxiety disorder; focused in response prevention, trying to reduce compulsions; focused in exposure to images, objects or situations that provoke obsessions; focused, finally, in changing old habits by new habits [14]. It is important to integrate symptomatology with patient's story: how the disorder appears, which factors are maintaining it, what kind of coping strategies the patient is using to fight with the obsessions and compulsions.

Third wave of CBT includes new themes: metacognition, cognitive fusion, emotions, acceptance, mindfulness, dialectics, spirituality, and therapeutic relationship [15]. Change in metacognitions is effective in treatment of OCD. Metacognition is the aspect of cognition that controls mental processes and thinking. To this model, the cognitive attentional syndrome, a psychopatological state consisting of repetitive cognitive processes such as worrying, rumination, dysfunctional threat monitoring, and dysfunctional cognitive and behavioral coping are at the core of depressive and anxiety disorders [15].

OCD patients reveal a significant lower proportion of positive emotions in dreams than other people, and their dreams tend to be shorter, less complex and less emotional. These characteristics are not modified after CBT treatment, except the trend to show less negative emotions during dreams. Exposure does not seem to have traumatizing effect [16].

In patients with anxiety there is a preferential encoding of information that is consistent with the treat-related concerns [5]. The objective of cognitive therapy is to correct dysfunctional beliefs and information processes biases.

4. Treatment of OCD from a cognitive constructive perspective

Usually cognitive behavioral treatment of OCD is directed to develop a less threatening explanation of world and to prevent compulsive responses to obsessive thoughts in order to neutralizing those behaviors.

During cognitive behavioural treatment some patients present an impairment of depressive symptoms, specifically during exposure, but This group presents a higher level of improvement after treatment. It must be due to the integration of new knowledge and experiences, which initially could lead to a bigger "chaos" [17]. Changes are associated to system destabilization, old patterns are less viable, and new patterns emerge, what provokes fluctuations.

Beliefs and behaviors are supposed to be related one to each other in OCD, and this relation can be bidirectional. During treatment, changes in beliefs can be followed by changes in conducts, but also changes in beliefs can be preceded by the behavioral changes [17].

Higher levels of predicted criticism are related to poorer treatment outcome, because negative feedback is found to be less effective if the individual feels he has no influence over his performance.

Attributional style involves communicating a causal belief about an event, and high levels of relatives' hostility is related to high levels of responsibility attributions. Cognitive treatment of OCD tries to develop a less threatening explanation for the world. From a constructive perspective, it's important to develop a less dangerous perceived world, in which personal control does not determine all the life circumstances.

Memory distortions may reflect the influence of adaptative processes that are beneficial for cognitive functions, but also result in memory errors [18]. These distortions are based on the operation of a schema that is useful to organize and interpret information. Episodic memory supports the construction of future events by extracting and recombining stored information into a simulation of a novel event. Brain activity is highly similar during remembering the past and imagining the future [18]. This similarity could explain the link between thought and act that is observed in some patients with OCD.

The elevated expressed emotions act as stressors for OCD patients and enhance symptomatology, so an objective of the treatment must be coping with those and / or family therapy to reduce them.

Neutralizing behaviors such as compulsions have as an objective to prevent some danger that would produce blame from significant ones. The OCD behaviors are maintained to avoid criticism but are converted to stereotypical behaviors that no longer function in terms of the original reasons. And OCD behavior itself provokes further criticism and acts as a stressor to maintain behaviors.

When criticism is a focus in OCD treatment, Some previous questions must be answered: who delivers criticism, how, where it is delivered, how does the recipient perceive the feedback, if it is followed by punishment, and if it is overlapped with blame. The therapy focus could be accepting or managing actual criticism when it is accurate and directed toward appropriate targets. If patient is in an unsupportive environment, therapeutic work must involve family or significant others who are the source of criticism [3].

Other cognitive mechanisms implicated in OCD are heightened responsibility, thought-action fusion (TAF), self-doubt, overimportance of thoughts, cognitive control, perfectionism, overestimation of threat, and intolerance of uncertainty [10]. Obsessions are viewed as interferences about reality, arrived at on the basis of an inductive narrative: the realty is perceived initially adaptively, but as a result of reasoning errors, that vision derives in obsessional interferences: he or she is influenced by self-generated narratives that lead them to doubt their perception of external experience in favor of a hypothetical, internally generated version. These errors produce inferential confusion when a remote possibility is conflated with a fictional narrative, so the previously imagined becomes a real possibility. Then the individual acts as if this imagined supposition is potentially real and is drive to try to modify it, albeit unsuccessfully [10].

OCD is related to "inflated responsibility", because responsibility is extended beyond its "normal" range [9]. A person with OCD acts "as if" personally responsible of the object of him or her obsession. They do not believe to have acted the obsession, but they feel as if it has happened. Some styles of information processing do not lend themselves to rational revision of beliefs, but the integration of these aspects of experience when integrating alternative narratives to the OCD narrative may change cognition in the process of therapy.

Imagination has a core role in OCD, as is demonstrated by the efficacy of imaginary exposure with response prevention. Imagining a future event increases the subjective likelihood that the event is going to occur, being more important when the event is ease to be thought. The concerns of OCD people are almost always about imagined events that have never occurred before, at least to them [10].

Treatment uses to be similar in children and adults, due to clinical similitude [19], but in younger there's more evidence in behavior psychotherapy and pharmacological than in a cognitive approach. In the cognitive therapy one of the components is relabeling obsessions as symptoms, not ordinary thoughts. Patients are encouraged to distance themselves from obsessions and view them as bizarre messages. Refocusing on an alternative behavior is important, too, as a way to tolerate response prevention. Group treatment for OCD can be useful due to the sense of "being in it together", that reduces anxiety [19].

Treatment of dysfunctional schemas is useful too in OCD, assuming these are in the origin of negative intrusions and its misinterpretations. Sometimes OCD is the result of the inability to cope with early experiences as abuse and emotional rejection. The assumption of a unitary and rational self underlies in cognitive view, though it has no restraining influence on the practice of cognitive therapy [9].

Narrative approach does not treat cognitions as stand-alone thought units, compared with veridical perception nor reduce them to schemas, but narrative theory considers thoughts in the experience and adaptative context of the person [9]. The narrative engages the person in his/her own problem and positions on it, so obsessions fit into a complete scene.

First stage in narrative therapy is to elicit narratives, asking people to present the problem in their own words, so it is easier to understand the relations among obsessional thoughts, personal relationships, beliefs... obsessions are not isolated items, empty of content. Thinking emerges as a part of a script, so responsibility can be origin of obsessions from "me" or from "others", as a way to understand oneself or something that affects relationships with other people through the possible errors.

In dialogical approach, the inner world is studied in the form of interpersonal relationships: the concept of one self is expanded to a multiplicity of relatively autonomous "I" positions, each one with one voice. The voices of separate selves are validated socially, are heard, silenced or modified through dialogue [9]. Obsessions have an alien nature, that could be understood as a thought or behavior designed as "not me", belonging to the unfamiliar self, revealed when people do not understand their own reactions. A dialogical approach sees the issue of designating and obsessional thought, image, or impulse as consistent or inconsistent with the "self" in a contextual and dialogical manner [9]. Some patients report that

http://dx.doi.org/10.5772/57579

they feel like another person when they are performing an obsessional ritual, they feel that the ritual is stronger than themselves. Self-generated narratives can modify confidence in alternative possibilities as well as actual perceptions.

The dialogical approach to obsessions interprets the specificity of obsessions in terms of an unresolved dialogue [9], that can be completed, rephrased or answered to move it to resolution. Many times the second voice is not present, so people with OCD are not able to answer it. But sometimes OCD people explain their obsessions as "voices", different from the reason voice. In any case, obsessional neutralizing behavior would be dictated by the power of narrative, exposing and identifying with the person the processes by which narratives are constructed.

With the schema theory, OCD emerges when a conflict appears between contradictory schemas. Development of a unitary self is not possible in presence of incompatible attachment schemas. An excessive sense of personal vulnerability threatens identity. But OCD patients can have a high functioning in many contexts, with the problem centered in a specific situation, where the obsession emerges.

Many obsessions are consistent with the two-stage model of intrusions, in the way "If I think about it, I will become very anxious", but some times there is a common framework, as to feel one is a sinner if a blasphemous thought (intrusion) appears. The themes of all obsessions fall into three categories, following Rachman (cited in 9]: aggression, sexuality and blasphemy. Intrusive thoughts become more frequent, persistent and unpleasant when the threat is bigger. It is not clear if the theme of an obsession is related to developmental experiences and prior beliefs. However, from narrative framework it is possible to link it to earlier development, so the content of the obsession can be traced to children-parents` previous experiences. The final aim of the treatment is to generate alternative accounts of the OCD experience which would modify the meaning of thoughts and behavior.

Viewing over-responsibility, exaggerated danger, improbable consequences and magical thinking within a narrative context helps to understand the development of them and explain the individual response pattern [9]. The therapy tries to empower the patient to modify his or her own obsessional narrative, not to ignore it or fight with it.

5. Obsessive-Compulsive Spectrum Disorders (OCSD)

Some disorders present clinical similarities with OCD, with the urge to execute a behavior though it is unwanted or lived as desadaptative followed by relief when it is done or anxiety if it could be done. These disorders are conceptualized as Obsessive-Compulsive Spectrum Disorders (OCSD). Some authors understand these as an addictive behavior. These disorders at the compulsive end respond better to SSRIs and those at the impulsive end (dysmorphic disorder, hypochondriasis, onychophagia, and psychogenic excoriation) appear to benefit from a wider range of thymoleptics [20]. But also all of them respond, at least partially, to CBT. Among Obsessive Compulsive Spectrum Disorders are described the following:

- Body Dysmorphic Disorder: associated to preoccupation with an imagined or overemphasized defect in appearance appear repetitive and often ritualistic behaviours (such as mirrorchecking and request for reassurance). Exposure to social situations avoiding camouflage, resist to compulsive behaviors and cognitive restructuring have shown to be beneficial. CBT plus psychopharmacologic treatment with Serotonin Reuptake Inhibitors (SRI) have demonstrated the higher response.
- Hypochondriasis: is the persistent fear or belief that one has a serious illness based of one's misinterpretation of body signs, which leads to hypervigilance. Treatment is based in restructuring faulty assumptions about physical symptoms and modifying maladaptative patterns of behavior that maintain symptomatology. Individual psychotherapy focusing on illness and symptom perception can be useful, too.
- Trichotillomania: recurrent pulling of one's hair with noticeable hair loss. There is an increased stress immediately before the behavior with posterior relief. Habit reversal therapy (selfmonitoring, competing response, thought stopping) is the more efficient CBT, also associated to acceptance and commitment therapy.
- Pathological gambling: uncontrollable urge or impulse to gamble that progressively increases in intensity until generating social and / or economic difficulties. Imaginary desensitization is more efficient than aversion relief or other behavioral treatments.
- Compulsive buying: implies shopping preoccupations or behaviors. The possible benefit of cognitive restructuring techniques to enable patients to develop more appropriated responses to their impulses has been suggested as efficient.
- Kleptomania: is characterized by the recurrent failure to resist impulse to steal items that are not needed for personal use or for its monetary value. Patients experience increased sense of tension prior to the act and a sense of pleasure, relief, or gratification when committing theft. There are no studies published about CBT in kleptomania.
- Onychophagia and psychogenic excoriation: chronic nail biting and compulsive skin picking are considered impulse control disorders, because the self-injurious behaviors are habitual, ritualistic, tension-reducing, and ego-dystonic. Competing response therapy is significantly superior to aversion therapy, and aversion therapy is superior to self-monitoring alone.

6. Conclusions

OCD is a chronic, severe, and sometimes incapacitating disorder which needs to be treated in a multifactorial way. Psychopharmacological treatment has demonstrated its utility in this illness, but the best results are obtained when it is associated to psychotherapy. This efficacy is more evident in patients who are resistant to pharmacotherapy alone.

One of the most proved therapies is cognitive behavioral therapy. CBT has many strategies to treat OCD. The most used strategies are exposure with response prevention, but these are not applicable in all the patients, because they can provoke an increase of anxiety which could be unbearable.

We propose in this revision to work with a narrative approach, putting the disorder in the frame of the patients' lives, understanding how and when it has emerged, trying to find the sense of it, to give a new look to the disorder and constructing a new way to cope with the irrational or overemphasized ideas.

There are not many evidences of constructive or narrative approach in treatment of OCD, but we think constructivism has an important role in OCD treatment. Understanding the beginning of the disorder and the role of it into patients' lives improve the way the patients can face the symptomatology. This perspective helps the patients to re-explain the symptoms as a way to cope with a fearsome world which could be adaptative in a moment but these symptoms have lost their coping function when they have transformed themselves in a rigid way.

Author details

Mª-José Martín-Vázquez*

Address all correspondence to: mjmv66@gmail.com

Psychiatry Department, Hospital Infanta Sofía, San Sebastián de los Reyes, Madrid, Spain

References

[1] Rodrigues, H; Figueira, I.; Gonçalves, R.; Mendlowicz, M.; Macedo, T.; Ventura, P. CBT for pharmacotherapy non-remitters—a systematic review of a next-step strategy *J Affect Disord*. 2011 Mar;129(1-3):219-28

[2] Ma, JD.; Wang, HC.; Li, HF.; Zhang, XL; Zhang, ZL; Hou, YH; Liu, XH, Hu, XZ, Cognitive-coping therapy for obsessive-compulsive disorder: A randomized controlled trial *Journal of Psychiatric Research* 47 (2013) 1785-1790

[3] Pace, S.M.; Thwaites, R.; Freeston, M.H. Exploring the role of external criticism in Obsessive Compulsive Disorder: a narrative review. *Clin. Psychol. Rev.*, 2011. 31: 361-370

[4] Borda, T.; Vázquez, F. Anormalidades del procesamiento de la información en el trastorno obsesivo-compulsivo. *Sinopsis*, 2008. 21 (40): 7-10

[5] Laposa, J.M.; Rector, N.A. Cognitive bias to symptom and obsessive belief threat cues in Obsessive-Compulsive Disorder. *J. Nerv. Ment. Dis.* 2009. 197: 599-605

[6] Schacter, D.L. Illusory memories: a cognitive neuroscience analysis. *Proc. Natl. Acad. Sci. USA*. 1996. 93: 13527-1533

[7] Buchanan, T.W. Retrieval of emotional memories. *Psychol. Bull.* 2007. 133 (5): 761-779

[8] Casado Martín, Y.; Cobos Álvarez, P.; Godoy Ávila, A.; Farias, A.; Vila Castellar, J. Procesamiento emocional en personas con sintomatología obsesivo-compulsiva. *Psycothema*, 2011. 23 (1): 94-99

[9] Hallam, R.S.; O'Connor, K.P. A dialogical approach to obsessions. *Psychol and Psychother: theory, research and practice.* 2002. 75: 333-348

[10] Keen, N.; Brown, G.P.; Wheatley, J. Obsessive compulsive symptoms and the simulation of future negative events. *Br. J. Clin. Psychol.*, 2008. 47, 265-279

[11] Brewin, C.R. Understanding cognitive behaviour therapy: a retrieval competition account. *Behav. Res. Ther.* 2006. 44: 765-784

[12] Asbahr, F.R.; Castillo, A.R.; Ito, L.M.; Latorre, R.D.; Moreira, M.N.; Lotufo-Neto, F. Group cognitive-behavioral therapy versus sertraline for the treatment of children and adolescents with Obsessive-Compulsive Disorder. *J. Am. Acad. Child Adolesc. Psychiatry*, 2005. 44 (11): 1128-1136

[13] Franklin, M.E.; Sapyta, J.; Freeman, J.B.; Khanna, M.; Compton, S. et al. Cognitive behavior therapy augmentation of pharmacotherapy in pediatric Obsessive-Compulsive Disorder. The pediatric OCD treatment study II (POTS II) randomized controlled trial. *JAMA*, 2011. 306 (11): 1224-1232

[14] Fuster Pérez, J. Una terapia integradora para las obsesiones. 9º congreso virtual de psiquiatría. Psiquiatria.com 2008

[15] Kahl, K.G.; Winter, L; Sxhweiger, U. The third wave of cognitive behavioural therapies: what is new and what is effective?. *Curr. Opin. Psychiatry*, 2012. 25: 522-528

[16] Kuelz, A.K.; Stotz, U.; Rienmann, D.; Schredl, M.; Voderholzer, U. Dream recall and dream content in Obsessive-compulsive patients: is there a change during exposure treatment?. *J. Nerv. Ment. Dis.* 2010. 198: 593-596

[17] Polman, A.; Bouman, T.K.; van Geert, P.L.C.; de Jong, P.J.; den Boer, J.A. Las creencias disfuncionales en el proceso de cambio del tratamiento cognitive en comprobadores obsesivo-compulsivos. *RET*, 2013.68: 3-19

[18] Schacter, D.L. Constructive memory: past and future. *Dialogues Clin. Neurosci.* 2012. 14: 7-18

[19] Söchting, I.; Third, B. Behavioral group treatment for obsessive-compulsive disorder in adolescence: a pilot study. *Int. J. Group Psychother.*, 2011. 61 (1): 85-97

[20] Ravindran, A.V.; da Silva, T.L.; Ravindran L.N.; Richter, M.A.; Rector, N.A. Obsessive-Compulsive Spectrum Disorders: a review of the evidence-based treatments. *Can. J Psychiatry*, 2009. 54 (5): 331-343

Chapter 6

Cognitive Behavioral Therapy for Obsessive Compulsive Disorder

Zena Al-Sharbati, Marwan Al-Sharbati and
Ishita Gupta

Additional information is available at the end of the chapter

http://dx.doi.org/10.5772/57332

1. Introduction

Obsessive Compulsive Disorder (OCD) is a neuropsychiatric condition characterized by obsessions (defined as recurrent and persistent thoughts, impulses or images that are intrusive and inappropriate and cause marked distress) and compulsions (repetitive behaviors or mental acts driven by obsessions to reduce distress or prevent dreaded events [1]. OCD is the fourth most prevalent psychiatric disorder and was considered as one of the most debilitating disorders with a prevalence rate of 2.5 % in an individuals' lifetime [2, 3]. OCD is widespread among both adults and children [4]. OCD is considered a relatively common occurrence in childhood and adolescence where prevalence rates are between 1 to 3.6% [5-7]. As it is the case with other mood and anxiety disorders, females tend to be at a higher risk of developing OCD compared to males [8]. About one third to one half of adult OCD cases has its origins in childhood or adolescence [9, 10]. However, a recent meta-analysis found that pregnant women and post partum women have higher prevalence rates of OCD compared to the general population, with postpartum women at a much greater risk compared to the general female population [11]. Research has found that early onset is associated with males while late onset is associated with females [10].

It is worthwhile to mention that the new DSM- 5 released in May, 2013 has a number of changes to OCD and related disorders such as hoarding disorder and Body Dismorphic Disorder (BDD). OCD is no longer an anxiety disorder, rather is has its own chapter in the DSM-5. The chapter includes other disorders, in addition to obsessive-compulsive disorder, such as Body Dysmorphic Disorder and trichotillomania (hair-pulling disorder), as well as two new disorders: hoarding disorder and excoriation (skin-picking) disorder [12].

The presence of co-morbidities is common in people with OCD. For example, some of the most common disorders associated with OCD are depression [13-15], anxiety [16-19], alcohol or substance misuse [20-22], BDD [23], or eating disorders [24]. People with OCD have a tendency for impulsive behavior (may include symptoms of childhood conduct disorder). They also show higher rates of suicidal attempts [25]. OCD is often considered a chronic condition and it significantly impairs various spheres of a person's life including social, emotional and somatic [25].

1.1. Etiology

The etiology of OCD is multifaceted. OCD is etiologically and phenomenologically heterogeneous [26]. For individuals with OCD, it has been indicated that there are abnormalities in the neurotransmitter; serotonin, which plays an important role in carrying nerve impulses from neurons to the receptor cells *via* the brain. The normal functioning of serotonin is impaired in patients with OCD as a result of blocked or impaired receptor sites in the brain [27]. Neurological aberrations could be associated with genetic mutations, while environmental factors can play a role in the expression of OCD symptoms [27]. The use of Clomipramine or Selective Serotonin Reuptake Inhibitors (SSRIs) has shown promising results in reducing symptoms of individuals with OCD [4, 28]. This indicates the involvement of the brain in the etiology of OCD [27, 28].

Although the same diagnostic criteria are used to diagnose OCD, its expression varies from person to person (i.e. onset, severity, obsessions and compulsions) [29]. Symptom dimensions are heritable in OCD [30]. It is imperative to pinpoint that the onset of OCD is linked to the interaction between both genetic and environmental correlates. This does not necessarily indicate that negative life events cause OCD, rather a number of studies have found that for people with biological or psychological vulnerability, negative events may trigger the onset of OCD [31, 32].

The identification of susceptible genes involved in the onset of OCD is challenging given the heterogeneity of the disorder. The use of symptom-based rather than disorder-based construct has reduced the heterogeneity of OCD [3]. vanGrootheest (2008) conducted a twin-study to explore the OC symptom dimension heritability [33]. The vanGrootheest study is pioneering in the field because it utilized population-based rather than OCD - sample of twin pairs. It identified rumination, contamination and checking; three OC symptom dimensions which shared discrepancy with OC behavior that was explained by both genetic (36%) and environmental factors (64%). However, certain genes influenced the contamination dimension and therefore, that dimension was independent. The use of family pedigrees is yet another approach to determine the heritability of the OCD trait [34]. In pedigree analysis, a high genetic loading was associated with symmetry and ordering symptoms. More interestingly, for symmetry, hoarding and contamination symptoms, correlations among independent sibling pairs were found [35, 36]. According to the OCD Collaborative Genetics Study, sibling-sibling association was associated with four factors; hoarding and taboo thoughts being the most prevalent [37, 38]. In families with a multigenerational history of OCD and hoarding, OC symptoms were found to have a higher rate of heritability compared to hoarding [39].

1.2. Symptoms and its Sub-types

Obsessive-Compulsive symptomatology is generally measured using the Yale-Brown Obsessive Compulsive Scale (Y-BOCS), which includes an assessment of the severity of the disorder and a checklist (YBOCS-CL) of the OC symptoms [40, 41]. This scale consists of 45 obsessions and 29 compulsions within 15 predefined symptom categories [40, 41]. A recent meta-analysis identified five OC symptom dimensions or factors: (i) symmetry and repeating; (ii) aggressive, sexual and religious obsessions; (iii) contamination and cleaning, (iv) aggressive obsessions and checking; and (v) somatic obsessions [42].

1.3. Diagnosis

According to the *DSM-IV* criteria, the symptoms of OCD should not be due to a general medical disease, chemical substance, or drug overdose [1]. OC symptoms are associated with several neurological lesions of the cortico-striatal-thalamic-corticol circuits resulting from the administration of drugs such as methylphenidate or cocaine (dopamine agonists) or after streptococcal infection [4]. Clinically speaking, OCD symptoms are associated with marked anguish and dysfunction. Subclinical OC symptoms are common and are noted during the path of normal development [4]. However, OCD can negatively affect a person's quality of life [43-45]. In some cases, diagnosis becomes unclear, when the symptoms are perplexed with those of Obsessive-Compulsive Personality Disorder [42]. Although, OCD symptoms vary from symptoms observed in other anxiety and mood disorders, they are an intrinsic component of other disorders such as autism and frontal lobe lesions [46]. Inappropriate diagnosis, under diagnosis and incompetent treatment are linked to tremendous direct and indirect costs [4].

2. Brief historical overview

Since the 4th century BC, OCD has been elucidated as "melancholia", a unique disease with certain mental and physical symptoms. These symptoms are attributed to the imbalance in the bodily fluids; blood, black bile, yellow bile, and phlegm [47]. Hippocrates defined all fears and depressions if lasting for a long time as symptoms of melancholia [47].

In the early Middle Ages in Europe, blasphemous, sexual or obsessive perceptions were attributed to satanic possession, which was believed to lead to compulsive behaviors [48]. Priests treated patients by expelling spirits from their bodies. However, in the 17th century, the concept of "religious melancholy" came to be known as the cause of several mental disorders including those with OCD-like symptoms [49].

Sigmund Freud, an Austrian psychiatrist, suggested that OC behavior is associated with unconscious conflicts [47, 50]. Freud's analytical conceptualization of "obsessive neurosis" was based on the idea that people develop OCD as a result of the difficulties in the anal psychosexual stage of development. Freud utilized psychoanalysis to treat OCD. There is no current evidence - based research that supports the use of traditional psychodynamic psychotherapy in the treatment of OCD [51]. This, in turn, gave rise to the behavioral mode of psychotherapy.

Several cognitive behavioral theories have been developed to understand and target symptoms' progression and maintenance in people with OCD. The following sections will discuss the various cognitive behavioral theories.

2.1. Mowrer's two stage theory

Also known as the two-factor theory, Hobart Mowrer proposed a strictly behavioral drive-based theory in 1947 to explain avoidance in entirely behavioral terms [52]. Mowrer's theory merged the learning principles of classical and operant conditioning and explains fear acquisition and the vicious cycle of rituals. Based on classical conditioning, phobias are the result of an association between a neutral stimulus (conditioned stimulus; CS) and feared stimulus (unconditioned stimulus; UCS) [53]. Mowrer's theory supports the concept that a CS provokes fear when it is recurrently presented with an UCS. The neutral stimulus is any mental event and/or physical object while the avoidance of the stimulus reduces anxiety [53]. People with OCD react to the neutral stimulus with the same fear level associated with the unpleasant event. Furthermore, it is through the generalization process that fear and avoidance can extend to other places or situations, which remind the individual of the trauma, triggering the same fear response. The persistence of behavioral avoidance leads to further distress. The theory also explains how rituals are maintained. Dollard and Miller (1950) further adopted Mowrer's theory to explain the progress and maintenance of fear/anxiety and avoidance in OCD [54].

In a series of experiments, Rachman and colleagues showed that obsessions increase obsessional distress while compulsions reduce distress levels [55, 56]. It was this concept that defined a functional relationship between obsessions and compulsions, defining OCD in *DSM-III* and its successors [1]. Recently, many studies explored the relationship between OCD and other anxiety disorders, and reconsidered the validity of the diagnostic classification of OCD [26, 57]. The core psychopathology of OCD is grounded on "preoccupation" and "repetitive behaviors", a result of cognitive and motor failure, which overlap with obsessive-compulsive and related disorders (OCRDs), recently included in DSM-5 and ICD-11 [58].

Further, Roper et al. (1973) noted that obsessions and compulsions were maintained to overcome associated anxiety. In addition, results showed that the urge to perform a particular activity disappeared within a 15-180 minute period if the patient did not engage in the activity [56]. The repetition of the exercises reduced the anxiety and urges to carry out these activities; the phenomenon being defined as "habituation" and formed the basis of EX/RP therapy. In 1970s, clinical trials added more support to the efficacy of exposure and ritual prevention (EX/RP) therapy in reducing OC symptoms, thus becoming one of the first choices for the treatment of OCD [59, 60].

2.1.1. EX/RP Therapy — Efficacy and limitations

In more than 70% of the cases, individuals treated with OCD responded to EX/RP therapy more efficiently compared to individuals who were on SSRIs [61]. However, the model has some limitations. About 20- 30% of individuals with OCD drop-out or refuse treatment. This is evident in cases with severe OCD or low insight level [62]. In addition, EX/RP had little or no benefit in helping people with pure obsessions. Furthermore, the model did not explain why OCD sufferers become distressed by aggressive or sexual thoughts, without having witnessed or committed such acts [63].

Cognitive theory and therapy was developed as a result of the limitations of the behavioral and ERP theories and therapies. Cognitive techniques focus on distorted thoughts and dysfunctional beliefs along with highlighting the role of cognitions in the origin and maintenance of symptoms [64]. The addition of cognitive techniques to ERP has been effective in the treatment of patients who presented with pure obsessions or obsessive ruminations [60]. Also, it led to the development of the "cognitive behavioral theories", which is now considered as the gold standard treatment for OCD [65].

2.2. Cognitive theories

Several cognitive theories emerged to understand the development and maintenance of OCD symptoms. According to Rachman and de Silva (1978), obsessions of a disturbing nature are a normal occurrence in people with and without OCD [66]. However, individuals with no OCD do not give importance to such thoughts and they, as a result, disappear easily, although they can play a critical role in transforming normal unpleasant thoughts into obsessions [66]. This section will focus primarily on two theories: Foa and Kozak's, and Salkovskis' theory.

2.2.1. Foa and Kozak's theory

According to Foa and Kozak, OCD is distinguished by two erroneous cognitions: (i) individuals with OCD have a cognitive vulnerability for the overestimation of the probability of danger in safe situations; (ii) they exaggerate the severity of the consequences of negative events [67].

2.2.2. Salkovskis theory

Salkovskis proposed an improved cognitive model in which responsibility was considered a critical issue for the onset of obsessions [68, 69]. He proposed five basic assumptions that are characteristic of individuals with OCD: (i) Thoughts and actions are the same; (ii) causing harm is the same as not preventing harm; (iii) despite difficult events, personal liability for harm continues; (iv) not engaging in harm related rituals is the same as having the intention to harm; (v) controlling one's thoughts is a personal obligation. Salkovskis's theory proposes that patients' catastrophic appraisals of obsessions lead to an increased sense of responsibility [70]. Such responsibility beliefs tend to cause increased anxiety and end with compulsions to exert control over obsessions [70]. Obsessions tend to persist as long as the distorted and blasphemous beliefs persist, and decline when such thoughts are enfeebled.

Rachman (2002) proposed a counterpart to this theory to explain the origins of compulsions [71]. According to Rachman (2002), scrutiny of rituals occurs as inflated thoughts of responsibility lead to compulsions aimed at reducing any possible damage. Scrutiny, as a result, leads to continuous scrutiny to eliminate doubt and possible risk [71].

However, several theorists suggested that 6 other belief domains could be associated with the origin and maintenance of OCD symptoms, including (i) inflated responsibility [68]; (ii) over-importance of thoughts [72]; (iii) extreme concern about the importance of controlling one's thought [73, 74]; (iv) overestimation of risk [75]; (v) intolerance of uncertainty [75]; and (vi) perfectionism [76].

Freeston et. al (1996) indicated various common features in patients who benefited from cognitive interventions and they include: (i) placing high value over the incidence of obsessions; (ii) believing that obsessions determine the patient's real character; (iii) equating thoughts and actions; and (iv) believing that the presence of a thought increase the possibility of its occurrence. Several cognitive techniques were added to ERP therapy, leading to the adoption of "Cognitive Behavioral Therapy (CBT)" [77].

Cognitive Therapy: Efficacy and Limitations: Cognitive therapy is efficient in the treatment of patients with predominant obsessions, and those with obsessions and compulsions [72]. Cognitive therapy can benefit patients with poor insight or those with dysfunctional beliefs, suggesting the use of cognitive therapy prior to EX/RP therapy to reduce anxiety and improve adherence to EX/RP exercises. Cognitive therapy has not shown to be effective in patients who significantly display obsessions of sexual, aggressive or destructive content. However, the use of cognitive therapy alone has not been established, giving rise to cognitive behavior therapy [73].

3. Cognitive Behavioral Therapy (CBT)

The CBT model for the treatment of OCD is brief and structured, and it aims to help patients cope with OCD symptoms. This is typically done through the utilization of behavioral as well as cognitive strategies. Some of these strategies involve exposure to fearful stimuli, response prevention, challenging dysfunctional erroneous beliefs, psycho-education, and home exercises such as symptom monitoring and exposure exercises [79].

When OCD symptoms are mild and are not associated with any co-morbidity, CBT is usually brief lasting between 13-20 sessions [80]. Initially, CBT is held on a weekly basis, or twice a week until some improvements are observed. As the condition improves, sessions are spaced. Behavioral exposure techniques help to achieve habituation to fear while cognitive techniques target dysfunctional beliefs, especially for patients with a preponderance for obsessions [47]. The following sections will describe the process of CBT, which include patient assessment followed by psycho-education, ERP and cognitive exercises, discharge and relapse prevention.

3.1. Patient assessment

A detailed psychiatric and psychological assessment is solicited from patients to rule out differential diagnosis and any associated disorders [81]. In some cases, specific brain imaging examinations may be included to rule out organic causes. The Yale-Brown Obsessive Compulsive Scale (*Y-BOCS*) can be utilized to acquire a comprehensive overview of the OCD symptoms. The effectiveness of CBT is reduced when OCD is accompanied by other co-morbid conditions such as severe anxiety or depression, psychosis and addiction [82].

3.2. Initiation of CBT

Psychoeducation: Once OCD diagnosis has been confirmed, the therapist utilizes psycho-education mini lessons to educate the patient about the nature of OCD, the CBT model of treatment, vicious maintenance cycles and the role of avoidance and fear in maintaining the disorder. In addition, patients will be educated about helpful and unhelpful coping strategies, escape and avoidance, exposure therapy, the development of fear hierarchy and principles of exposure. Given the crucial role of homework, patients and their families are introduced to the role of homework within the therapy framework [83]. Engaging family members in this stage is crucial, given the challenge family members face in changing their daily routines and responding to questions with reassurances to accommodate the patient [84].

Motivation: Despite the growing body of research supporting the use of EX/RP in the treatment of OCD, it is estimated that about 25% of patients with OCD refuse EX/RP [85]. This refusal may be associated with the increased exposure to anxiety, which is an intrinsic part of the treatment [86]. Lack or low level of motivation is especially problematic for pediatric patients with OCD, and is associated with negative therapeutic outcomes. Therefore, interweaving motivational interviewing techniques within the therapy framework helps patients decrease their ambivalence about treatment. A recent study found that using the motivational interviewing within therapy sessions proved helpful in facilitating speedy improvement and minimizing the burden for families [87].

3.2.1. Elaboration of list of hierarchy

The initial task in CBT is to work with patients to create a list of anxiety provoking situations that targets main OCD symptoms. The Y-BOCS-CL is frequently used by clinicians to create a comprehensive list. Anxiety provoking situations will be ranked by patient according to the level of anxiety it creates starting from 1 (mild anxiety) to 4 (high level of anxiety). To increase the patient's self-efficacy, exposure exercises start with the least anxiety provoking situation [88].

Assessment measures: To monitor the progress of patients throughout the therapeutic process, it is crucial to utilize assessment measures at various points during treatment. One of the widely used sensitive scales is the Yale-Brown Scale, which is a valid and reliable instrument for assessing the severity of obsessive-compulsive disorder [40, 41]. The Yale-Brown Scale is composed of 10 questions; 5 to assess obsessions and 5 to assess compulsions. The scale assesses severity from (0- no symptom, to 4 –severe symptoms). A maximum of 40 points can be obtained for patients with severe symptoms [40, 41]. Another sensitive tool is the revised version of the Obsessive Compulsive Inventory. Its usefulness has been proven as a diagnostic tool for screening patients with OCD [89].

3.2.2. Exposure and response prevention exercises (EX/RP)

In Vivo exposure (involves the direct confrontation with the fearful stimulus) is the most common form of exposure. Utilizing the already developed hierarchy list, the patient along with the therapist initiate therapy by choosing a situation that evokes a medium level of anxiety. Therapy sessions are focused on generalizing and extending gains to other situations through designing homework exercises. It is crucial to discuss with the patient that while confronting anxiety may increase anxiety, over time, anxiety will dissipates. Response prevention increases anxiety level, but it gradually declines. That is why response prevention reduces the need to engage in compulsions [61]. The main effect of EX/RP exercises is the instant increase of anxiety in the session, which reaches its peak during the first exercise and decreases or even disappears during the next 15-180 minutes [90].

3.3. Cognitive exercises

The initiation of cognitive exercises occurs when the patient have mastered EX/RP exercises and is content with identifying OC symptoms. The patient should be able to make a distinction between obsessions and normal thoughts. In these exercises, various parameters including unpleasant thoughts, their background and emotive, physiological and behavioral concerns are documented. Patients are taught the classifications of these beliefs based on their domains. Cognitive techniques include the recognition and documentation of unpleasant thoughts; Socratic questioning; downward arrow; and pie of responsibility [69].

3.4. Ending CBT and relapse prevention

Once OCD symptoms have significantly improved, spacing sessions are proposed and CBT is ceased. Given that OCD is a chronic disorder with high relapse rates [91], both therapists and patients work to prevent relapse through creating relapse prevention plans. Booster sessions help in monitoring the maintenance of therapeutic gains [92].

4. Cognitive behavioral psychotherapy adaptation for children and adolescents with obsessive compulsive disorder

Childhood OCD is linked with severe commotion in social and academic functioning, co-morbid emotional and behavioral complications and family difficulties [93]. Cognitive-Behavioral Therapy (CBT), especially exposure and response prevention (ERP) is the pivotal therapy used for OCD in adults [94]. Exposure encompasses useful confrontation of objects or situations that trigger obsessions; response prevention includes refraining from activities that reduce anxiety generated by obsessions [93]. Initially, ERP was considered more helpful for adults compared to children and adolescents.

However, since the mid-1990s, research proved ERP to be viable, safe and effective for both children and adults [95]. However, not many children and adolescents receive CBT for various factors, one being lack of trained clinicians. Interestingly, clinicians find the research-driven-treatment methodology not practical [93]. Furthermore, even though OCD in children is more prevalent, variations in dimensions between adults and children (age, maturity, language development..etc) obfuscate the application of CBT for children. Children are not able to recognize or label their obsessions or fear responses. In certain cases, a child's responses tend to mislead adults into believing that the child's behavior is willful [96]. One of the challenges facing the child therapist is children's lack of understanding of OCD symptoms. High anxiety levels during ERP makes children resistant to engage in therapy. EX/RP homework exercises are challenging to children given the aversion towards homework. This in turn makes parental supervision crucial for a successful treatment [93]. Criticism and conflicts within the family negatively affect therapy [97]. Dropout rates are associated with the rigidity in applying treatment protocols, suggesting the need to assess developmental issues, and consequently adopting treatment plans tailored to meet each child's needs [93].

Recently, CBT protocols have been developed for children. These protocols typically include psycho-education in age-appropriate language, and cognitive strategies to cope with anxiety, and the use of reinforcement contingent strategies within the family environment [98]. Parents' active involvement in the child's treatment increases the effectiveness and long-term gains from the treatment [99]. A sensitive protocol for CBT in children was recently developed and is considered a flexible and practical guide for clinicians; RIDE Up and Down the Worry Hill [100, 101].

4.1. RIDE up and down the worry hill

The abbreviation RIDE and the metaphor of riding a bicycle Up and Down the Worry Hill were designed to describe basic CBT principles in a child-friendly manner [100, 102].

The "Worry Hill" explains the relationship between exposure and habituation. It explains the gradual increase in anxiety levels as a result of exposure to the fearful stimuli. Anxiety levels continue to increase as exposure to fear continues. The fear will reach the top peak in the Worry Hill. This is when autonomic habituation starts and anxiety declines. However, if the child surrenders to or escapes fear, habituation is interrupted and obsessions are strengthened [102].

The four-step RIDE acronym (Rename, Insist, Defy, Enjoy) comprises the steps that either the child or adolescent should take to complete the Worry Hill [100, 102]. The RIDE was developed to streamline EX/RP for children and adolescents and raise persistence of anxiety until habituation occurs. RIDE includes cognitive (externalizing, distancing) and behavioral (Controlling obsessions, exposure and self-reinforcement) techniques. Coaching is provided in each of the four steps followed by therapist modeling, behavioral rehearsal, repeated practice and reinforcement, until the individual has grasped each step [102]. The following section will discuss each of the four steps.

4.1.1. R: Rename the thought (That's OCD talking, not me!!)

This is the first step and comprises recognition of OCD thoughts as separate and different from the child's rational self [101, 102]. The method of externalizing OCD has been employed by

Schwartz (1996) [103] with adults and March at el (1994) [98] with children and has been found to be effective.

4.1.2. I: Insist that YOU are in charge (I'm in charge. I choose NOT to believe OCD!)

The second step allows the child to use the power of choice [101, 102]. Insisting that oneself is in charge helps build self-confidence and strength needed to embark on exposure.

4.1.3. D: Defy OCD-Do the OPPOSITE of what it wants (I will RIDE up the worry hill and stick it out until i can coast down)

This step involves ERP, specifically a change in behavior [101, 102]. Response prevention involves exposure to the obsession and associated fear response without engaging in compulsions. This escalates anxiety till it reaches the peak, and once habituation is set, the anxiety starts to decline. With continuous practice, anxiety habituates at a faster rate.

4.1.4. E: ENJOY your success — Reward yourself (I did it! I BEAT OCD! I can do it again)

This is the final step and lets the child review his/her success and take due credit for effort [101, 102].

The Worry Hill represents a universal metaphor as children, adolescents and even adults can relate to the conception of riding a bicycle up a hill and makes it easier to understand the principle behind CBT. The implementation of CBT in both children and adolescents occurs in four sequential phases, which will be discussed in the following section.

4.2. Implementation of CBT in children and adolescents

The four phases are focused on specific goals and aims to develop various skills in every phase. The severity of the disorder determines the number of psychotherapy sessions. In some cases, a total of 6 sessions are required, whereas in others, an average of 10-20 sessions is required.

4.2.1. Phase 1 — Biophyschosocial assessment and treatment plan

This phase lays the foundation for effective treatment and takes from one to three sessions, each session equaling 50-minutes. This assessment is designed to be culturally sensitive and comprehensive to include an appreciation of OCD symptoms in a holistic manner. This involves joint efforts amongst the therapist, parent, child and other concerned personnel. The aim is to develop an individualized child-centered treatment plan. The use of clinical interviews, self-report inventories and behavioral observations are important at this stage [101].

4.2.2. Phase 2 — Building treatment readiness

This phase emphasizes planned and active preparation for treatment. Although a critical phase, it tends to be ignored, leading to frequent treatment failures. This phase requires one to three sessions and includes stabilization of any family crisis and the use of effective communication skills to increase awareness about treatment and builds therapeutic rapport [101].

4.2.3. Phase 3 — The RIDE up and down the worry hill

This phase comprises of 4 to 15 sessions and consists of separate as well as joint sessions with both child and parents. It is during this stage that the child participates in EX/RP [101].

4.2.4. Phase 4 — After the RIDE

As the final step of the treatment, it is initiated when the child and parents have mastered the RIDE and the RALLY effectively, respectively as well as when there is a decline in the OCD symptoms [101].

5. Efficacy studies of the cognitive behavioral psychotherapy and pharmacology for OCD

In a systematic review, Foa and Kozak (1996) found that about 83% of patients with OCD included in trials (n=330), showed a minimum of 30% improvement in their symptoms compared to pre-treatment measures. These trials utilized exposure and response prevention techniques [104].

When exploring long-term outcomes of patients with OCD who utilized CBT, a review of 17 studies indicated that about 76% of patients continue to show improvements in their OCD symptoms at an average of about two years post treatment [104]. In addition, CBT was shown its effectiveness not only in the treatment of OCD but also other mental health disorders such as depression and generalized anxiety disorder as evidenced by a review of 16 meta-analyses [105].

On the other hand, Gava et.al (2007) conducted a systematic review of eight randomized trials comparing the therapeutic outcomes of patients receiving a form of cognitive behavioral therapy versus patients who were treated as usual [106]. Treatment as usual included no treatment, waiting list or usual care. Results from 214 patients indicated that cognitive-behvaioral therapy is associated with more significant decline in OCD symptoms compared to treatment as usual [106]. Amelioration of 50-70 % of OCD symptoms were noted for patients with OCD involved in the above mentioned trials. Therapeutic outcomes did not differ among the variations of cognitive behavioral therapy. In addition, results showed no difference in the clinical outcomes for patients with OCD who were assigned to individual or group psychotherapy sessions.

Various randomized clinical studies were conducted to control for confounding variables such as the presence of co-morbid disorders, OCD severity, age, just to name a few. However, Franklin et al. (2000) compared the results of 4 randomized controlled trials (EX/RP) with treatment outcomes for 110 clinical adult patients receiving the same treatment modality on a paid outpatient service [107]. Results showed that clinical patients showed significant improvement in their OCD symptoms that were comparable to patients in randomized trials [107].

Many factors have been identified to help reach favorable clinical outcomes [108]. In session (in vivo exposure) have been found to be associated with better clinical outcomes compared to exposure techniques done by patient alone at home [108]. Tremendous reductions in symptoms is associated with complete response prevention, as opposed to partial response prevention [108].

CBT is indicated for patients with OCD with various levels of symptom severity. It is worthwhile to mention some of the considerations when evaluating a patient's suitability for this treatment modality which include

1. An understanding of the treatment module and rationale, given that the implementation of CBT requires a good grasp of main concepts that are practiced by the patient as homework. Therefore, the utilization of CBT with people who are cognitively challenged may be difficult.

2. Low insight level for the OCD symptoms poses a challenge for patients and therapists. Low insight is associated with reduced engagement in exposure techniques, thus rendering the use of cognitive techniques essential to increase insight levels for patients [109]. In addition, individuals with predominantly hoarding symptoms have showed lower responses to CBT [104].

3. Patient's preference: CBT is preferred by patients with no co-morbid conditions, mild to moderate OC symptoms, or those with a preference for psychotherapy over medications [110]. Furthermore, CBT is the preferred choice for pregnant women [110]. CBT is highly effective for patients who did not or partially respond to treatment with psychotropic medications [111, 112]. However, in some cases both medications and CBT are preferred [113]. However, some cases warrant the use of medications such as the presence of severe OCD symptoms, low psychological insight, and co-morbidities [113]. Respecting the patient's choice of his/her preferred treatment modality, after discussing the pros and cons of each, is pivotal and is linked to increased adherence to treatment [114].

4. The presence of co morbid conditions such as depression and anxiety along with OCD reduces the efficacy of CBT techniques [108].

5. Negative emotional involvement by family members was found to be associated with higher treatment dropout rates. Hostility was a consistent factor associated with early dropout rates and negative treatment outcome [115].

Other Psychotherapy Modalities: Among all psychosocial interventions and techniques, exposure and response prevention, has led to better clinical outcomes for patients with OCD [116]. Given the absence of randomized clinical trials to investigate the effectiveness of other psychotherapeutic modalities in psychotherapy [106], CBT is considered as the gold *standard for the treatment of OCD.*

6. Psychotropic medications or CBT?

The utilization of CBT is associated with better clinical outcomes compared to the use of Serotonin Reuptake Inhibitors. Although Clomipramine (CMI) was considered as the first line drug therapy associated with significant improvements in OCD symptoms, [117], the randomized trial conducted by Foa et al. (2005) showed that clinical outcomes did not differ for the two control groups (EX/RP alone, and EX/RP plus Clomipramine) [61]. In addition, research has indicated that D-cycloserine has shown to produce favourable clinical outcomes when augmented with CBT [118]. Specifically, D-cycloserine helped to reduce response times after exposure exercises [119].

7. Conclusion

Over the last 40 years, EX/RP has been recommended as first choice treatment for OCD, with CBT as an alternate. More work is needed to enhance CBT programs and make it sensitive to the population needs. Future studies need to explore the treatment responses for specific sub-types of OCD patients such as "washers" and "checkers". Such understanding will help in designing sensitive treatment plans for the various sub-types of OCD. Also, future studies need to incorporate patients' populations with co-morbid conditions as they tend to be excluded from such studies, thus making it difficult to know the efficacy level for CBT in this patient population.

Abbreviations

OCD	Obsessive Compulsive Disorder
CBT	Cognitive Behavioural Therapy
DSM	Diagnostic and Statistical Manual of Mental Disorders
BDD	Body Dysmorphic Disorder
SSRIs	Selective Serotonin Reuptake Inhibitors
YBOCS	Yale Brown Obsessive Compulsive Scale
YBOCS-CL	Yale Brown Obsessive Compulsive Scale Checklist
CS	Conditioned Stimulus
UCS	Unconditioned Stimulus
ICD	International Classification of Disorders
EX/RP	Exposure and Response Prevention
CMI	Clomipramine

Author details

Zena Al-Sharbati[1*], Marwan Al-Sharbati[2] and Ishita Gupta[3]

*Address all correspondence to: zenamarwan@yahoo.com

1 Department of Behavioural Medicine, Sultan Qaboos University Hospital, Sultan Qaboos University, Muscat, Oman

2 Department of Behavioural Medicine, Sultan Qaboos University, Oman

3 Department of Genetics, Sultan Qaboos University, Oman

All authors declare "no conflict" of interest.

References

[1] American Psychiatric Association. Diagnostic and Statistical Manual of Mental Disorders.4th ed. Washington, DC: American Psychiatric Association; 1980.

[2] Karno M, Goldin JM, Sorenson SB et al. The epidemiology of obsessive-compulsive disorder in five US communities. Archives of General Psychiatry, 1988; 45: 1094-99.

[3] Katerberg H, Delucchi KL, Stewart SE, Lochner C, Denys D, Stack DE, Andresen JM, Grant JE, Kim SW, Williams KA, den Boer JA, van Balkom A, Smit JH, van Oppen P, Polman A, Jenike MA, Stein DJ, Mathews CA, Cath DC. Symptom Dimensions in OCD: Item-Level Factor Analysis and Heritability Estimates. Behav Genet, 2010; 40(4): 505-517.

[4] Stein DJ. Obsessive-compulsive disorder. The Lancet. 2002; 360: 397-405.

[5] Flament MF, Whitaker A, Rapoport JL, Davies M, Berg CZ, Kalikow K, Sceery W, Shaffer D. Obsessive compulsive disorder in adolescence:An epidemiological study. Journal of the American Academy of Child and Adolescent Psychiatry, 1988; 27: 764–771.

[6] Valleni-Basile LA, Garrison CZ, Jackson KL, Waller JL, McKeown RE, Addy CL, Cuffe SP. Frequency of obsessive-compulsive disorder in a community sample of young adolescents.Journal of the American Academy of Child and Adolescent Psychiatry, 1994; 33: 782–791

[7] Zohar H, Ratzoni G, Pauls DL, Apter A, Bleich A, Kron S, Rappaport M, Weizman A, Cohen DJ. An epidemiological study of obsessive-compulsive disorder and related disorders in Israeli adolescents. Journal of the American Academy of Child and Adolescent Psychiatry, 1992; 31: 1057–1061.

[8] Pigott TA. Obsessive-compulsive disorder: symptom overview and epidemiology. Bull Menninger Clin, 1998 ;62(4 Suppl A): A4-32.

[9] Kolada JL, Bland RC, Newman SC. Obsessive-compulsive disorder. Acta Psychiatrica Scandinavica Supplementum, 1994; 376: 24–35.

[10] Noshirvani HF, Kasvikis Y, Marks IM, Tsakirisn F, Monteiro WO. Gender-divergent aetiological factors in obsessive-compulsive disorder.British Journal of Psychiatry, 1991; 158: 260–263.

[11] Russell EJ, Fawcett JM, Mazmanian D. (2013). Risk of obsessive-compulsive disorder in pregnant and postpartum women: a meta-analysis. Journal of Clinical Psychiatry, 2013; 74 (4):377-85.

[12] Grohol, J. DSM-5 changes, Obsessive –Compulsive and Related Disorders. Psych Central, 2013. http://pro.psychcentral.com/2013/dsm-5-changes-obsessive-compulsive-and-related disorders/004404.html (Accessed on September 11 2013).

[13] Abramowitz, J.S. Treatment of obsessive-compulsive disorder in patients who have comorbid major depression. Journal of Clinical Psychology, 2004; 60, 1133–1141.

[14] Abramowitz, J.S., Schwartz, S.A., Moore, K.M. et al. Obsessive-compulsive symptoms in pregnancy and the puerperium: a review of the literature. Journal of Anxiety Disorders, 2003; 17: 461–478.

[15] Apter A, Horesh N, Gothelf D, et al. Depression and suicidal behavior in adolescent inpatients with obsessive compulsive disorder. Journal of Affective Disorders, 2003; 75: 181–189.

[16] Biederman J, Petty C, Faraone SV, et al. Moderating effects of major depression on patterns of comorbidity in patients with panic disorder. Psychiatry Research, 2004; 126, 143–149.

[17] LaSalle, V.H., Cromer, K.R., Nelson, K.N., et al. Diagnostic interview assessed neuropsychiatric disorder comorbidity in 334 individuals with obsessive compulsive disorder. Depression and Anxiety, 2004; 19, 163–173.

[18] Nestadt, G., Addington, A., Samuels, J., et al. The identification of OCD:related subgroups based on comorbidity. Biological Psychiatry, 2003; 53, 914–920.

[19] Welkowitz, L.A., Struening, E.L., Pittman, J., et al. Obsessive-compulsive disorder and comorbid anxiety problems in a national anxiety screening sample. Journal of Anxiety Disorders, 2000; 14, 471–482.

[20] Abram KM, Teplin LA, McClelland GM, *et al.* Comorbid psychiatric disorders in youth in juvenile detention. American Journal of Psychiatry, 2003; 60:1097–1108.

[21] Bakken K, Landheim AS, Vaglum P. Primary and secondary substance misusers: do they differ in substance-induced and substance-independent mental disorders? Alcohol and Alcoholism, 2003; 38, 54–59.

[22] Fals-Stewart W, Angarano K. Obsessive-compulsive disorder amongpatients entering substance abuse treatment: prevalence and accuracy ofdiagnosis. Journal of Nervous and Mental Disease, 1994; 182, 715–719.

[23] Frare F, Perugi G, Ruffolo G, *et al.* Obsessive-compulsive disorder and body dysmorphic disorder: a comparison of clinical features. European Psychiatry, 2004; 19,292–298.

[24] Jordan J, Joyce PR, Carter FA, *et al.* Anxiety and psychoactive substanceuse disorder comorbidity in anorexia nervosa or depression. International Journal of Eating Disorders, 2003; 34, 211–219.

[25] Hollander E, Greenwald S, Neville D, Johnson J, Hornig CD, Weissman MM. Uncomplicated and comorbid obsessive-compulsive disorder in an epidemiological sample. Depress Anxiety, 1997; 4: 111-19.

[26] Matsunaga H. Current and emerging features of obsessive-compulsive disordertrends for the revision of DSM-5.SeishinShinkeigakuZasshi, 2012; 114(9): 1023-30.

[27] Aouizerate B, Guehl D, Cuny E, Rougier A, Bioulac B, Tignol J, Burbaud P. Pathophysiology of obsessive-compulsive disorder: a necessary link between phenomenology, neuropsychology, imagery and physiology. Prog Neurobiol, 2004; 72(3): 195-221.

[28] The Clomipramine Collaborative Study Group. Clomipramine in the treatment of patients with obsessive-compulsive disorder. Arch Gen Psychiatry, 1991; 48(8): 730-8.

[29] Mataix-Cols D, Pertusa A, Leckman JF. Issues for DSM-V: how should obsessive-compulsive and related disorders be classified? American Journal of Psychiatry, 2007; 164:1313–1314.

[30] Pauls DL. The genetics of obsessive compulsive disorder: a review of the evidence. Am J Med Genet C Semin Med Genet, 2008; 148C:133–139.

[31] Gothelf D, Aharonovsky O, Horesh N, et al. (2004) Life events and personality factors in children and adolescents with obsessive-compulsive disorder and other anxiety disorders. Comprehensive Psychiatry, 45, 192–198.

[32] Khanna S, Rajendra PN, Channabasavanna SM. Life events and onset of obsessive compulsive disorder. International Journal of Social Psychiatry, (1988b); 34: 305–309.

[33] vanGrootheest DS, Boomsma DI, Hettema JM et al. Heritability of obsessive-compulsive symptom dimensions. Am J Med Genet B Neuropsychiatr Genet. 2008; 147B: 473–478.

[34] Alsobrook IJ, Leckman JF, Goodman WK et al. Segregation analysis of obsessive-compulsive disorder using symptom based factor scores. Am J Med Genet, 1999; 88: 669–675.

[35] Cullen B, Brown CH, Riddle MA et al. (2007). Factor analysis of the Yale-Brown obsessive compulsive scale in a family study of obsessive-compulsive disorder. Depress Anxiety;24:130–138.

[36] Chacon P, Rosario-Campos MC, Pauls DL et al. Obsessive compulsive symptoms in sibling pairs concordant for obsessive compulsive disorder. Am J Med Genet B Neuropsychiatr Genet. 2007; 144B:551–555.

[37] Hasler G, Pinto A, Greenberg BD et al. Familiality of factor analysis-derived YBOCS dimensions in OCD-affected sibling pairs from the OCD Collaborative Genetics Study. Biol Psychiatry, 2007; 61:617–625.

[38] Pinto A, Greenberg BD, Grados MA et al. Further development of YBOCS dimensions in the OCD collaborative genetics study: symptoms vs. categories. Psychiatry Res, 2008;160:83–93.

[39] Mathews CA, Nievergelt CM, Azzam A et al. Heritability and clinical features of multigenerational families with obsessive compulsive disorder and hoarding. Am J Med Genet B Neuropsychiatry Genetics, 2007; 144:174–182.

[40] Goodman WK, Price LH, Rasmussen SA, Mazure C, Fleischmann RL, Hill CL, Heninger GR, Charney DS. The Yale-Brown Obsessive Compulsive Scale. I. Development, use, and reliability. Arch Gen Psychiatry, 1989; 46(11):1006-11.

[41] Goodman WK, Price LH, Rasmussen SA, Mazure C, Delgado P, Heninger GR, Charney DS. The Yale-Brown Obsessive Compulsive Scale. II. Validity. Archives of General Psychiatry, 1989; 46(11): 1012-6.

[42] Bloch MH, Landeros-Weisenberger A, Rosario MC, Pittenger C, Leckman JF. Meta-analysis of the symptom structure of obsessive-compulsive disorder. American Journal of Psychiatry, 2008; 165(12): 1532-42.

[43] Leon AC, Portera L, Weissman MM. The social costs of anxiety disorders. British Journal of Psychiatry, 1995; 166 (Suppl 27), 19–22.

[44] Antony MM, Roth D, Swinson RP, et al. Illness intrusiveness in individuals with panic disorder, obsessive-compulsive disorder, or social phobia. Journal of Nervous and Mental Disease, 1998; 186, 311–315.

[45] Koran, L.M. Quality of life in obsessive-compulsive disorder. Psychiatric Clinics of North America, 2000; 23, 509–517.

[46] Hollander E. Obsessive-compulsive related disorders. Washington, USA: American Psychiatric Press, 1993.

[47] Allison A. A Brief History of Obsessive Compulsive Disorder: Ideas about OCD, a serious mental illness, have evolved from ancient notions of melancholia and possession through to Freud's psychoanalysis and modern neurochemistry. Melancholia,

2013; 1514. http://suite101.com/article/brief-history-of-obsessive-compulsive-disorder-a70602. (Accessed on 29th August 2013).

[48] Aardema F, O'Connor K. The menace within: obsessions and the self. International Journal of Cognitive Therapy, 2007; 21: 182-197.

[49] Berrios GE. Obsessive Compulsive Disorder: Its conceptual history in France during the 19th Century. Comprehensive Psychiatry, 1989; 30(4): 283-95.

[50] Cordioli AV, Vivan A. Cognitive-Behavioral Therapy of Obsessive-CompulsiveDisorder, Standard and Innovative Strategies in Cognitive Behavior Therapy, Dr. Irismar Reis De Oliveira (Ed.) 2012. ISBN: 978-953-51-0312-7, InTech, Available from: http://www.intechopen.com/books/standard-and-innovativestrategies-in-cognitive-behavior-therapy/cognitive-behavioral-therapy-of-obsessive-compulsive-disorder. (Accessed on 28th August 2013).

[51] Jenike MA. Obsessive-compulsive disorder: efficacy of specific treatments as assessed by controlled trials. Psychopharmacology Bulletin, 1993; 29(4): 487-99.

[52] Mowrer OH. Stimulus response theory of anxiety.Psychology Review, 1939; 46:553–565.

[53] Foa EB. Cognitive behavioral therapy of obsessive-compulsive disorder.Dialogues in Clinical Neuroscience, 2010; 12(2): 199-207.

[54] Dollard J, Miller NE. Personality and Psychotherapy; an Analysis in Terms of Learning, Thinking, and Culture. New York, NY:McGraw-Hill, 1950.

[55] Roper G, Rachman S. Obsessional-compulsive checking: experimental replication and development. Behav Res Ther. 1976; 14:25–32.

[56] Roper G, Rachman S, Hodgson R. An experiment on obsessional checking. Behav Res Ther. 1973; 11(3): 271–277.

[57] Stein DJ, Fineberg NA, Joseph Bienvenu O, Denys D́, Lochner C, Nestadt G, Leckman JF, Rauch SL, Phillips KA. Should OCD be classified as an Anxiety Disorder in DSM-V? Depression and Anxiety, 2010; 27: 495-506.

[58] American Psychiatric Association. Diagnostic and Statistical Manual of Mental Disorders (Fifth ed.). Arlington, VA: American Psychiatric Publishing, 2013, pp. 5–25.

[59] Marks I, Hodgson, R, Rachman, S. Treatment of chronic obsessive-compulsive neurosis by in-vivo exposure – a two year follow-up and issues in treatment. The British Journal of Psychiatry, 1975; 127: 349-364.

[60] Foa E, Goldstein A. Continuous exposure and complete response prevention in the treatment of obsessive-compulsive neurosis. Behavioral Therapy, 1978; 9(5): 821-829.

[61] Foa E, Liebowitz M, Kozak M, Davies S, Campeas R, Franklin M, Huppert J,Kjernisted K, Rowan V, SchmidtA, Simpson H, Tu X. Randomized, placebo-controlled trial of exposure and ritual prevention, clomipramine, and their combination in the treat-

ment of obsessive-compulsive disorder.The American Journal of Psychiatry, 2005; 162(1); 151-161.

[62] Stanley MA, Turner SM. Current status of pharmacological and behavioral treatment of obsessive-compulsive disorder.Behavior Therapy, 1995; 26, 163–186.

[63] Antony M, Purdon C, Summerfeldt L. Psychological treatment of obsessive compulsive disorder: Fundamentals and Beyond. American Psychological Association. Washington D.C., USA, 2007.

[64] Wright JH, Basco MR, Thase ME. Learning cognitive-behavior therapy. An illustrated guide. American psychiatric publishing Inc: Washington DC, 2006.

[65] Roth A, Fonagy P. What works for whom? A critical review of psychotherapy Research (2nd ed.). London: Guilford Publications, Inc, 2005.

[66] Rachman D, de Silva P. Abnormal and normal obsession.Behaviour Research and-Therapy. 1978; 16(4): 233-248.

[67] Foa EB, Kozak MJ. Treatment of anxiety disorders: implications for psychopathology. In: Tuma AH, Maser JD, eds. Anxiety and the Anxiety Disorders. Hillsdale, NJ: Lawrence Erlbaum Associates, 1985; 421–452.

[68] Salkovskis P. Obsessional-compulsive problems: a cognitive-behavioral analysis. Behavior Research and Therapy. 1985; 23(5): 571-583.

[69] Salkovskis P, Forrester E, Richards C. Cognitive-behavioural approach to understanding obsessional thinking. The British Journal of Psychiatry. 1998; 173(35): 53-63.

[70] O'kearney R. Responsibility Appraisals and Obsessive-Compulsive Disorder: A Critique of Salkovski'S Cognitive Theory. Australian Journal of Psychology.1998; 50(1): 43-47.

[71] Rachman S. A cognitive theory of compulsive checking. *Behaviour Research and Therapy,* 2002; 40(6): 625-639.

[72] Frost RO, Steketee G. (Eds.). Cognitive approaches to obsessions and compulsions: Theory,assessment, and treatment. Amsterdam: Pergamon Press, 2002.

[73] Purdon C, Clark DA. The need to control thoughts. In R. O. Frost & G. Steketee (Eds.), Cognitive approaches to obsessions and compulsions:Theory, assessment, and treatment (pp. 29–44). Amsterdam: Pergamon Press, 2002.

[74] Thordarson DS, Shafran R. Importance of thoughts. In R. O. Frost & G. Steketee (Eds.), Cognitive approaches to obsessions and compulsions: Theory, assessment, and treatment (pp.15–28). Amsterdam: Pergamon Press, 2002.

[75] Sookman D, Pinard G. Overestimation of threat and intolerance of uncertainty in obsessive compulsive disorder. In R. O. Frost & G. Steketee (Eds.), Cognitive ap-

proaches to obsessions andcompulsions: Theory, assessment, and treatment (pp. 63–90). Amsterdam: Pergamon Press, 2002.

[76] Frost RO, Novara K, Rhéaume J. Perfectionism in obsessive compulsive disorder.In R O. Frost & G. Steketee (Eds.), Cognitive approaches to obsessions and compulsions: Theory,assessment, and treatment (pp. 91–106). Amsterdam: Pergamon Press, 2002.

[77] Freeston M, Rhéaume J, Ladouceur R. Correcting faulty appraisal of obsessional thoughts. Behavior Research and Therapy, 1996; 34(5-6): 433-446.

[78] Abramowitz J, Taylor S, McKay D. Obsessive-compulsive disorder. Lancet, 2009; 374(9688): 491-499.

[79] Whittal M. Land, McLean PD. CBT for OCD: The rationale, protocol and challenges. Cognitive behavioral practice, 1999; 6: 383-396.

[80] Koran L, Hanna G, Hollander E, Nestadt G. Practice guideline for the treatment of patients with obsessive-compulsive disorder. The American Journal of Psychiatry, 2007; 164(7): 5–53.

[81] Swedo SE, Leonard HL, Gravy M et al. Pediatric autoimmune neuropsychiatric disorders associated with streptococcal infections: clinical description of the first 50 cases. American Journal of Psychiatry.1998; 155: 264-71.

[82] Raffin A, Fachel J, Ferrão Y, de Souza F, Cordioli A. Predictors of response to group cognitive-behavioral therapy in the treatment of obsessive-compulsive disorder. European Psychiatry, 2009; 24(5): 297-306.

[83] Houghton S, Saxon D. An evaluation of large group CBT psycho-education for anxiety disorders delivered in routine practice. Patient education and counseling, 2007; 68: 107-110.

[84] Calvocoressi L, Lewis B, Hariis M, Trufan S, Goodman W, McDougle C, Price L. Family accommodation in obsessive-compulsive disorder.The AmericanJournal of Psychiatry, 1995;152(3): 441-443.

[85] Franklin ME, Foa EB. Cognitive behavioral treatments for obsessive-compulsive disorder. In J. M. Gorman (Ed.), A guide to treatments that work (pp. 339–357); New York: Oxford University Press, 1998.

[86] Maltby N, Tolin DF. A Brief Motivational Intervention for Treatment-Refusing OCD Patients.Cognitive Behavior Therapy, 2005; 34(3): 176-184.

[87] Merlo LJ, Storch EA, Lehmkuhl HD, Jacob ML, Murphy TK, Goodman WK, Geffken GR. Cognitive Behavioral Therapy Plus Motivational Interviewing Improves Outcome for Pediatric Obsessive–Compulsive Disorder: A Preliminary Study. Cognitive Behaviour Therapy, 2010; 39(1), 24-27.

[88] Salkovskis PM, Richards C, Forrester E. Psychological treatment of refractory obsessive compulsive disorder and related problems, in Obsessive Compulsive Disorder:

contemporary issues in treatment (personality and clinical psychology series). Edited by Goodman, WK, Rudorfer, M.V., Mahwah, N.J, Lawrence Earlbaum Associates, Inc, 2000, pp.201-221.

[89] Foa E, Huppert J, Leiberg S, Langner R, Kichic R, Hajcak G, Salkovskis P. The Obsessive Compulsive Inventory: Development and validation of a short version. The Psychological Assessment, 2002; 14(4): 485-495.

[90] Hupper JD, Roth DA. Treating obsessive-compulsive disorder with exposure and response prevention. The behavior Analysis today, 2003; 4(4): 66-70.

[91] Braga D, Cordioli A, Niederauer K, Manfro G. Cognitive-behavioral group therapy for obsessive-compulsive disorder: a 1-year follow-up. ActaPsychiatricaScandinavica, 2005; 112(3): 180-186.

[92] Bell J. Relapse Prevention in the Treatment of OCD. International OCD Foundation, 2012. http://www.ocfoundation.org/eo_relapse.aspx. (Accessed on 01 September 2013).

[93] Wagner AP. Cognitive-Behavioral Therapy for Children and Adolescents with Obsessive-Compulsive Disorder.Brief Treatment and Crisis Intervention, 2003; 3(3): 291-306.

[94] Marks, I. Behavior therapy for obsessive compulsive disorder. A decade of progress.Canadian Journal of Psychiatry, 1997; 42: 1021–1026.

[95] March J, Franklin M, Nelson A, Foa, E. Cognitive-behavioral psychotherapy for pediatric obsessive-compulsive disorder.Journal of Clinical Child Psychology, 2001; 30: 8–18.

[96] Rapoport J, Inoff -Germain G, Weissman M, Greenwald S, Narrow W, Jensen P, et al. Childhood obsessive-compulsive disorder in the NIMH MECA study. Parent vs. child identification of cases.Journal of Anxiety Disorders, 2000; 14: 535–548.

[97] Hibbs E, Hamburger S, Lenane M. Determinants of expressed emotion in families of children and adolescents. Journal of Child Psychology and Psychiatry, 1991; 32: 757–770.

[98] March J, Mulle K, Herbel B. Behavioral psychotherapy for children and adolescents with obsessive-compulsive disorder: An open trial of a new protocol-driven package. Journal of the American Academy of Child and Adolescent Psychiatry, 1994; 33: 333–341.

[99] Waters T, Barrett P, March JS. Cognitive-Behavioral family treatment of childhood obsessive-compulsive disorder.American Journal of Psychotherapy, 2001; 55(3): 372–387.

[100] Wagner AP. What to do when your child has obsessive-compulsive disorder: Strategies and solutions, 2002; Rochester, NY: Lighthouse Press.

[101] Wagner AP. Treatment of OCD in children and adolescents: A cognitive-behavioral therapy manual, 2003; Rochester, NY: Lighthouse Press.

[102] Wagner AP. Up and down the Worry Hill: A children's book about obsessive-compulsive disorder, 2000; Rochester, NY: Lighthouse Press.

[103] Schwartz JM. Brain lock. New York: Harper-Collins, 1996.

[104] Foa EB, Kozak MJ. Psychological treatment for obsessive-compulsive disorder. In: Long-term treatments of anxiety disorders, Mavissakalian MR, Prien RF (Eds), American Psychiatric Press, Inc, Washington, DC 1996. p.285.

[105] Butler AC, Chapman JE, Forman EM, Beck AT. The empirical status of cognitive-behavioral therapy: a review of meta analyses. Clin Psychol Rev, 2006; 26(1): 17-31.

[106] Gava I, Barbui C, Aguglia E, et al. Psychological treatments versus treatment as usual for obsessive compulsive disorder (OCD). Cochrane Database Syst Rev, 2007; CD005333.

[107] Franklin ME, Abramowitz JS, Kozak MJ, Levitt JT, Foa EB. Effectiveness of exposure and ritual prevention for obsessive-compulsive disorder: randomized compared with nonrandomized samples. J Consult Clin Psychol, 2000; 68(4): 594-602.

[108] Abramowitz JS, Foa EB. Does comorbid major depressive disorder influence outcome of exposure and response prevention for OCD? Behav Ther, 2000; 31:795.

[109] Foa EB, Abramowitz JS, Franklin ME, Kozak MJ. Feared consequences, fixity of belief, and treatment outcome in patients with obsessive-compulsive disorder. Behav Ther, 1999; 30:717.

[110] Flosnik DL, Khin EK. Cognitive-Behavioral Therapy for Obsessive-Compulsive Disorder in Pregnancy.Psychiatric Annals, 2012; 42(7): 269-271.

[111] Tolin, D. Maltby, N., Diefenbach, G., Hannan, S., Worhunsky, P. (2004). Cognitive behavioral therapy for medication nonresponders with obsessive-compulsive disorder: a wait-list-controlles open trial. Journal of Clinical Psychiatry, 2004; 65(7): 922-931.

[112] Tenneij, N., van Megen, H, Denys, D., Westenberg, H. (2005). Behavior therapy augments response of patients with obsessive-compulsive disorder responding to drug treatment. Journal of Clinical Psychiatry, 2005; 66(9): 1169-.1175.

[113] Stein DJ, Denys D, Gloster AT, Hollander E, Leckman JF, Rauch SL, Phillips KA. Obsessive-compulsive disorder: diagnostic and treatment issues. Psychiatr Clin North Am, 2009; 32(3): 65-85.

[114] Deacon BJ, Abramowitz JS. Patient's perception of pharmacological and cognitive-behavioral treatments for anxiety disorders. Beh Ther, 2005; 36:139.

[115] oa EB. Failure in treating obsessive-compulsives. Behav Res Ther,1979; 17:169.

[116] Abramowitz JS. Effectiveness of psychological and pharmacological treatments for obsessive-compulsive disorder: a quantitative review. J Consult Clin Psychol, 1997; 65:44.

[117] Fernandez-Cordoba, E., Lopez-IborAlino J. Monochlorimipramine in mental patients resisting other forms of treatment. ActasLusoEspNeurolPsiquiatr, 1967; 26:119–147.

[118] Chasson GS, Buhlmann U, Tolin DF, et al. Need for speed: evaluating slopes of OCD recovery in behavior therapy enhanced with d-cycloserine. Behav Res Ther 2010; 48:675.

[119] Kushner MG, Kim SW, Donahue C, et al. D-cycloserine augmented exposure therapy for obsessive-compulsive disorder. Biol Psychiatry 2007; 62:835.

Chapter 7

Treatment-Resistant Obsessive-Compulsive Disorder (OCD) – Current Understanding

Mushtaq A. Margoob, Rajesh Chandel,
Huda Mushtaq and Dhuha Mushtaq

Additional information is available at the end of the chapter

http://dx.doi.org/10.5772/57232

1. Introduction

Obsessive-compulsive disorder (OCD) is a chronic condition represented by a diverse group of symptoms that include intrusive thoughts, rituals, preoccupations, and compulsions. These recurrent obsessions or compulsions cause severe distress to the person. The obsessions or compulsions are time-consuming and interfere significantly with the person's normal routine, occupational functioning, usual social activities, or relationships [1].

In the 10th revision of International Statistical Classification of Diseases and Related Health Problems (ICD-10), OCD is diagnosed when either obsessions or compulsions (or both) are present for a period of at least 2 weeks, acknowledged as excessive or unreasonable and are repetitive and unpleasant. The patient tries to resist them and acknowledges that they are originating in the mind and are not imposed by outside persons or influences as well as the obsessive thought or the compulsive act is not experienced as pleasurable. [2] Recently American Psychiatric Association's (APA) Diagnostic and Statistical Manual of Mental Disorders, DSM-5 separated it from anxiety disorder group and reclassified it with disorders like Body Dysmorphic Disorder, Hoarding Disorder, Trichotillomania, and excoriation disorder [3].

A few decades ago OCD had been regarded as a psychiatric condition that was mostly treatment-refractory [4]. Modern pharmacotherapy for OCD began with the observation that colmipramine but not other TCAs, relieved OC symptoms. This led to hypothesis that serotonin might be involved in pathophysiology of OCD5. Also numerous investigations showing the therapeutic action of different selective serotonin reuptake inhibitors (SSRIs) in OCD, and by additional findings, such as the provocation of OCD symptoms by the serotonergic agent m-chlorophenylpiperazine the role of serotonergic system in OCD was further made strong. Not only is serotonin uptake blockade an apparent prerequisite for clinical improvement, but there is a correlation between the magnitude of clinical response and the reduction in various serotonin markers during treatment with these drugs [6]. There is substantial evidence that suggests OCD patients may benefit from addition of antipsychotics to their ongoing SSRI treatment, suggesting that dopamine also might play a role in the pathophysiology of OCD [7]. Latest neuroimaging techniques have shown various structural and volumetric abnormalities in prefrontal and basal ganglia region in patients suffering from OCD. In prefrontal cortex there may be dopaminergic hyperfunction and in basal ganglia there may be hypofunction of serotonergic system. This dysfunction of the 'cortico-striato-thalamic' loops is strongly linked to the symptoms of OCD, where the dopamine has an inhibitory effect on GABA neurons. The ascending serotonergic projections from the raphe nuclei restrain and control the function of these loops. Thus, when serotonergic hypofunction is present, the predominantly dopaminergic loops became overactive, which has been confirmed by neuro-imaging techniques and by neurocognitive tests as well. This could be the reason for the fact that SSRIs have limited success in the treatment of OCD symptoms [8]. The role of glutamate, which is the primary excitatory neurotransmitter, is being increasingly investigated in the pathophisiology of OCD [9]. Glutamate is major excitatory neurotransmitter in the cortico-striatal-pallido-thalamocortical circuits and is also known to interact extensively with serotonin and dopamine [10]. Moore et al. (1998) reported striking changes in caudate glutamatergic concentrations (Glx)-a marker for combined glutamate and glutamine, resonance on proton magnetic resonance spectroscopy (1H-MRS) in a paediatric OCD patient following treatment with Paroxetine [11]. Rosenberg et al. (2000) studied 11 psychotropic drug-naïve children with OCD, with single-voxel 1H-MRS examinations and demonstrated that caudate Glx concentrations were significantly greater in the patients compared to healthy controls [12]. They also found that the caudate Glx levels in patients decreased significantly following 12-weeks treatment with Paroxetine to levels comparable to that of controls and the decrease was associated with decrease in symptom severity of the OCD patients, while there was no difference in Glx levels in the occipital cortex between the two groups. Whiteside et al. (2006) have demonstrated that within the right orbitofrontal white matter, relative levels of Glx and N-acetylaspartate were increased in adult patients with OCD compared with healthy controls and greater levels of Glx/Creatine were associated with more severe OCD symptoms. [13] Evidences from genetic studies are also suggesting the role of glutamate in pathophsiology of OCD. Investigators have reported statistically significant association between OCD and a locus on chromosome 9p24 that codes for a high-affinity neuronal/epithelial excitatory amino acid transporter (EAAC-1), also known as SLC1A1 (Solute carrier family 1, member 1) [14, 15]. It is hypothesized that in the brain this transporter is crucial in terminating the action of the excitatory neurotransmitter glutamate and in maintaining extracellular glutamate concentrations within a normal range [16].Researchers have examined the role of the opioid system and neuropeptides like oxytocin and vasopressin in OCD although the evidence has been equivocal [17, 18, 19].

About 40-60% of obsessive compulsive disorder (OCD) sufferers do not respond to appropriate courses of treatment with serotonin reuptake inhibitors (SRI) and even to combination of medicines [20]. Those OCD patients who do not achieve a satisfactory response after an adequate trial of first line therapies are described as treatment resistant. The satisfactory response is usually defined by a reduction in the Yale-Brown Obsessive Compulsive scale score ≥35% or ≥25% with respect to baseline [21].

The first selective serotonin reuptake inhibitor (SSRI), fluoxetine, was introduced in 1987. This was followed by other SSRI's i.e.fluvoxamine, paroxetine, sertraline, citalopram and escitalopram. The SSRIs are each structurally and chemically distinct. Escitalopram, an isomer of citalopram, is the only exception. This molecular diversity explains why individual responses to, and tolerability of, SSRIs are so varied. Fluoxetine has the longest half-life: 4 to 6 days. The half-life of sertraline is 26 hours, 35 hours for citalopram, 27 to 32 hours for escitalopram, 21 hours for paroxetine, and 15 hours for fluvoxamine. Sertraline, fluoxetine, and paroxetine being the most highly bound and escitalopram being the least bound. All SSRIs are metabolized in the liver by the cytochrome P450 enzymes. The most important drug-drug interactions involving the SSRIs occur as a result of the SSRIs inhibiting the metabolism of a co administered medication [1]. Citalopram and escitalopram are the most selective inhibitors of serotonin reuptake, with very little inhibition of norepinephrine or dopamine reuptake and very low affinities for histamine H_1, GABA or benzodiazepine receptors. The other SSRIs have a similar profile, except that fluoxetine weakly inhibits norepinephrine reuptake and binds to 5-HT_{2C} receptors; sertraline weakly inhibits norepinephrine and dopamine reuptake; and paroxetine has significant anticholinergic activity at higher dosages and binds to nitric oxide synthase [1].

According to a large number of published randomized controlled trials (RCTs), meta-analyses, current expert guidelines, and consensus statements, SSRIs and clomipramine are recommended as first-line agents for drug treatment of OCD [22, 23, 24, 25]. Guidelines of the World Federation of Societies of Biological Psychiatry (WFSBP) for the pharmacological treatment of OCD grant the highest category of evidence ("A", ie, full evidence from several RCTs) for the SSRIs escitalopram, fluvoxamine, fluoxetine, paroxetine, and sertraline, as well as for the TCA clomipramine. Since clomipramine is less well tolerated than the SSRIs, it was given a recommendation grade of 2 (moderate risk benefit ratio), while the SSRIs received the highest recommendation grade 1 (good risk: benefit ratio). For citalopram a recommendation grade of 3 (limited evidence from controlled studies) was given as only one positive double-blind, placebo-controlled study was published. According to WFSBP guidelines usually lower response rates are achieved in OCD in comparison with other anxiety disorders, and that sometimes only partial remission is achieved and higher doses are required for these drugs in OCD than for other anxiety disorders. In a systematic review of all long-term, placebo-controlled trials with SSRIs in OCD, the likelihood of relapse during 24 to 52 weeks of treatment was significantly lower on an SSRI than with placebo [26].

According to the British National Institute for Health and Clinical Excellence (NICE) and the Royal College of Psychiatrists guidelines, the initial pharmacological treatment in adults with OCD should be one of the following SSRIs: fluoxetine, fluvoxamine, paroxetine, sertraline, or citalopram [27]. A Cochrane review of placebo-controlled SSRI trials in OCD, comprising 17 studies with 3097 participants showed efficacy for all SSRIs (citalopram, fluoxetine, fluvoxamine, paroxetine, and sertraline) but no statistical differences in short-term therapeutic action among the individual SSRIs [28].

2. Treatment of OCD patients refractory to serotonergic antidepressants

Although SSRIs and clomipramine have proven efficacy in OCD, about 40% to 60% of patients show no or just partial symptom improvement to a treatment with a first-line drug [29]. The large fraction of patients without substantial response to standard treatment experiences significant morbidity [30]. Response is typically operationalized as a decline in symptoms, as measured by a 30–35% reduction in the Yale-Brown Obsessive-Compulsive Scale (Y-BOCS) [31, 32] or a decline in symptoms below a threshold of 16 (i.e., the boundary between mild and moderate OCD symptoms). So those patients who are "treatment responsive" as per operationalized criterion may continue to have levels of symptoms in the mild-moderate range and spend hours daily preoccupied with their obsessions and compulsions. We need novel treatment modalities for these partially responsive as well as treatment resistant patients with OCD.

High-dose treatment with serotonergic drugs is one of the available strategy which should be considered first. Greater improvement with higher vs lower doses of SSRI was reported using 250 to400 mg/d vs 200 mg/d of sertraline [33]and with escitalopram after an increase of dose from 20 up to 50 mg/d [34]. In a Open-label study of high (30 mg) and moderate (20 mg) dose escitalopram for the treatment of obsessive-compulsive disorder by Dougherty et al found that the 30 mg (vs. 20 mg) dose of escitalopram may provide a superior reduction in OCD symptoms for those sufferers with comorbid depression and/or anxiety [35].

Studies with tricyclic antidepressants, which act on both serotonin and norepinephrine, suggested that blockade of serotonin reuptake was required for antiobsessional effect [36]. In a meta-analysis of studies by Geller et al., including over 1000 adolescents with OCD clomipramine was shown to have greater efficacy compared with fluoxetine, fluvoxamine, paroxetine, and sertraline [37]. Clomipramine, which retains some norepinephrine reuptake inhibitory (NRI) effect relative to the SSRIs and its metabolite desmethylclomipramine has significant NRI activity, appears to be more effective than the more selective agents in meta analyses, though not in head-to-head comparisons leads to the hypothesis that dual-acting agent might be more effective in treatment- resistant patients [38]. An open-label trial of venlafaxine, which is a dual acting agent, both in treatment- naive and treatment-resistant OCD, suggested superior efficacy to SSRIs [39], In a head-to-head comparison between venlafaxine and clomipramine suggested comparable efficacy to clomipramine. [40]

Another strategy is to use other agents that target the serotonin system in a different way. Mirtazapine, a 5HT2A agonist, can accelerate the effect of paroxetine on OCD symptoms [41]. But the long-term benefit of mirtazepine and citalopram augumentation over citalopram alone is unclear [42]. Mirtazapine has also been shown to be efficacious as monotherapy in OCD patients in an open trial [43].

Intravenous clomipramine [44, 45] and intravenous citalopram [46] have been shown to lead to a substantial improvement in symptoms in some treatment-resistant patients.

http://dx.doi.org/10.5772/57232

3. Augmentation with other drugs

A number of different agents are being used as augmentation strategies for standard SRI therapy in treatment- refractory OCD, with variable success. Agents with modes of action beyond the serotonin system have shown the capacity to significantly improve symptoms in patients with limited response to SRI therapy alone [47].

Tourette's syndrome is highly comorbid with Obsessive-compulsive disorder in the pediatric population [48]. Overlapping pathological changes in the basal ganglia are implicated in the two disorders [49, 50]. As we see both typical and atypical antipsychotics have positive effects in Tourette's syndrometrials of dopamine antagonists, augmentation therapy in OCD might be worth considering [51, 52].

The combination of the antipsychotics risperidone, haloperidol, olanzapine, or quetiapine with an SSRI was shown to be more effective than SSRI monotherapy in treatment-resistant cases and is recommended (grade 3, ie, limited evidence from controlled studies) by the WFSBP guidelines [25]. Antipsychotics should be added after at least 3 months of maximally tolerated therapy of an SSRI. Response usually commences within one month duration as seen in most of the studies and about one third of treatment-refractory OCD patients show a clinically meaningful improvement. Superior effects of quetiapine versus ziprasidone as an adjunct to SSRI were found in treatment-resistant OCD patients in a retrospective study [53]. A single-blind study comparing risperidone versus olanzapine augmentation of SSRIs showed positive responses without differences between the two treatment groups [54]. In a 12-week, open-label, trial of aripiprazole, significant improvement of OCD symptoms was demonstrated [55]. In a small, open-label study suggest that aripiprazole as monotherapy was helpful for treating OCD [56].Augmentation of paroxetine with perospirone resulted in improvement in a case report [57].

A smaller number of studies have examined the utility of augmentation therapy with opioid agents. Morphine given once weekly has shown efficacy in treatment-resistant obsessive-compulsive disorder in a single doubleblind study, and administration of the opioid agonist tramadol hydrochloride has also been shown to diminish OCD symptoms [58, 59, 60].The mechanism of this interesting effect is unknown, but one of the proposed mechanism being opioids inhibiting glutamate release in cortex through disinhibition of serotonergic neurons.

Other augmentation agents include clonazepam, inositol, clonidine, monoamine oxidase inhibitors, and antiandrogens [61].

Positive results have been reported with longterm augmentation with citalopram (up to 60 mg/d) in 20 treatment-resistent OCD patients on clomipramine [62].

Riluzole is an antiglutamatergic agent that is approved by the Food and Drug Administration for neuroprotection in amyotrophic lateral sclerosis (ALS) [63]. Various proposed mechanisms of action of riluzole are inhibition of sodium currents in glutamatergic axon terminals, reducing neurotransmitter release [64]; reduction of P/Qtype calcium currents in the axon terminals, with a similar effect on glutamate release [65]; extension of the open time of Kv1.4 potassium channels [66]; and increased astrocytic uptake of glutamate [67]. A case report and an open label study suggest augmentation with the antiglutamatergic agent riluzole in SRI-resistant OCD patients has significant efficacy [68, 69].

The amino acid N-acetylcysteine (NAC) is widely used for its antioxidant properties and as an antidote for acetaminophen toxicity. Preclinical studies suggest that NAC also modulates CNS glutamate. NAC is converted to cystine, a substrate for the glutamate/ cystine antiporter located on glial cells. The uptake of cystine by glia causes glial release of glutamate into the extrasynaptic space, where it appears to stimulate inhibitory metabotropic glutamate receptors on glutamatergic nerve terminals and thereby reduces the synaptic release of glutamate [70].There is evidence for benefit of NAC augmentation to SSRI's in treating SRI-refractory OCD in two preliminary case reports [71, 72].

In an open-label augmentation trial with memantine, an N-methyl-Daspartate (NMDA) glutamate receptor antagonist, improvement of symptoms was seen in about half of the patients, who had failed to respond to treatment with an SSRI for at least 3 months [73]. Feusner et al in 2009 in an open label trial, compared the differential efficacy of memantine for obsessive-compulsive disorder vs. generalized anxiety disorder. Ten OCD and 7 GAD subjects received 12 weeks of open-label memantine 10 mg twice daily, as either monotherapy or augmentation of their existing medication. Primary outcome measures were the Yale-Brown Obsessive Compulsive Scale (YBOCS) for the OCD group, the Hamilton Anxiety Rating Scale (HARS) for the GAD group, and the Clinical Global Impression-Improvement Scale (CGII) for both groups. The results suggest that memantine may have preferential efficacy in the treatment of OCD versus GAD. But further larger placebo controlled studies are required [74].A Single-Blinded Case-Control Study of Memantine in Severe Obsessive-Compulsive Disorder provides preliminary supportive evidence for the effectiveness of memantine as a glutamatergic augmenting agent in severe OCD [75].

Adjunctive glycine (an NMDA glutamate receptor agonist) was also tested in a small double-blind placebo-controlled trial and approached efficacy for treatment of OCD symptoms but with high dropout rates [76].

Amantadine (another NMDA antagonist) could be a useful drug for the treatment of OCD according to preclinical findings [77].

Augmentation of SSRI's with topiramate, in treatment-resistant OCD patients may be beneficial [78, 79]. Double-blind studies with topiramate are being conducted [80].

Pregabalin, which can indirectly inhibit glutamate release via blockade of calcium channels, beneficial effects on OCD symptoms in combination with serotonergic antidepressants have been reported in case reports [81, 82].A double-blind placebo-controlled study with pregablin in SSRI-refractory OCD is being conducted [83].

D-cycloserine (a glutamatergic partial N-methyl-d-aspartate (NMDA) agonist) augmentation of psychotherapy with exposure and response prevention in OCD has so far been investigated in three randomized, double-blind, placebo-controlled studies. Significantly greater decreases in obsession-related distress after four exposure sessions under D-cycloserine (125 mg, given

2 hours before each session) were reported, however, after additional sessions, the placebo group tended to catch up [84]. In another study, OCD patients were reported to be significantly more improved under D-cycloserine at mid-treatment (ten behavior therapy sessions in total, dose of 100 mg 1 hour before each session), but not at later time points, and the D-cycloserine group's depressive symptoms were significantly more improved at posttreatment [85].

In a randomized, double-blinded, placebo-controlled augmentation trial examining CBT + D cycloserine versus CBT + Placebo with 30 youth (aged 8–17) with a primary diagnosis of OCD received seven exposure and response prevention sessions paired with DCS or placebo taken 1 hour before sessions. Compared with the CBT + Placebo group, youth in the CBT + D cycloserine arm showed small-to-moderate treatment effects. These results provide initial support for a more extensive study of D Cycloserine augmentation of CBT among youth with OCD [86].

It has been reported that canna-binoids inhibit glutamate release in the CNS. As discussed above glutamate is implicated in pathophisiology of OCD. Addition-ally, cannabinoid type 1 (CB1) receptors are distributed abun-dantly in the striatum, a brain region frequently associated with OCD [87, 88].

Two case reports showed Improvement in Refractory Obsessive Compulsive Disorder With cannabinoid Dronabinol [89].

4. Augmentation with or switch to cognitive-behavioral psychotherapy

In a RCT, patients who were put on adequate dosages of SSRI for at least 12 weeks, and in those who continued to have clinically significant symptoms, the augmentative effect of exposure and ritual prevention versus stress management training was compared; Group with exposure and response prevention had significant decrease of symptom severity of at least 25% by the end of 8 weeks [90].In a controlled open trial, patients with a history of an inadequate response to multiple SSRI medications in adequate doses were treated with 15 sessions of outpatient CBT. OCD symptoms decreased significantly and gains were maintained over 6 months [91]. A meta-analysis of psychotherapy and pharmacotherapy for OCD found highest effect sizes for combined treatment [92].

5. Invasive treatment options

5.1. Psychosurgery

Treatment resistant obsessive-compulsive disorder is one of the few diagnoses in modern psychiatry for which invasive neurosurgical procedures are part of the established treatment options. All ablative neurosurgical techniques target the CSTC circuits that are believed to be hyperactive in OCD. Anterior cingulotomy involves a lesion targeting the anterior cingulated cortex and cingulum. Anterior capsulotomy targets the subcaudate white matter, interrupting frontothalamic fibers. Limbic leucotomy combines these two, lesioning both cingulum and subcaudate white matter [93].

5.2. Deep brain stimulation

Several small-scale controlled and open studies have suggested that deep brain stimulation (DBS) of the internal capsule and/or the adjacent ventral striatal region may be of benefit to severely affected OCD patients who have exhausted conventional therapies [94, 95, 96, 97, 98, 99, 100, 101].

Benjamin et al in 2006 found promising long-term effects of DBS in highly treatment-resistant OCD [102].

5.3. Electroconvulsive therapy (ECT)

Maletzky and colleagues in a retrospective study of 32 patients with treatmentresistant OCD who received ECT between 1979 and 1991 (19 non-depressed and 13 depressed; 14 were primarily checkers, 13 primarily cleaners, 4 did both). They were treated with bilateral frontotemporal ECT and were evaluated 2 days prior to the start of treatment and at 5 days, 6 months, and 12 months after treatment. Five days after treatment, there were highly significant differences between the pre- and post-scores (paired t-tests; $P<.001$). At 6-months posttreatment, the differences were still significant. In this study, ECT had an anti-obsessional effect and the improvements occurred equally frequently and to an equal extent in depressed and non-depressed groups [103]. In addition, several single-case reports provided further evidence of the possible efficacy of ECT in treatment-resistant OCD. However, the unblinded design of these reports, the frequent comorbidity with other Axis I psychopathology (schizophrenia, depression, and Tourette's syndrome), and the different parameters of ECT used limit the confidence that can be placed in the findings [104, 105, [106, 107, 108].

5.4. Repetitive transcranial magnetic stimulation (rTMS)

Greenberg and colleagues randomized 12 patients with OCD into a singleblind trial. Patients were treated with 1 session of active right-side, active left-side, or sham (occipital position) TMS. Based on previous reports of prefrontal hypermetabolism or hyperperfusion in patients with OCD,44 Greenberg and colleagues believed that the application of a highfrequency rTMS to prefrontal cortex would have transiently interrupted OCD symptoms. The main finding was that right prefrontal rTMS had modest acute effects on compulsions whereas effects of left lateral prefrontal and occipital stimulation were transient and nonsignificant. Obsessions appeared unaffected by rTMS. Treatment was well-tolerated, with two patients reporting mild headache after stimulation [109]. Sachdev and colleagues randomized 12 treatment-resistant OCD patients to a 10-session, single-blind, 2-week trial of active right-side or left-side TMS. Ten subjects were taking medication (benzodiazepines, antidepressants, or antippsychotics) and had been maintained on a constant dose for 8 weeks prior to and during the period of the study. Evaluations after 2 weeks of stimulation and 1 month after the completion of the treatment showed significant reduction in obsessions and compulsions in both groups with no significant difference between right and left stimulation on the overall Y-BOCS score. The authors suggested that rTMS may be beneficial in the acute treatment of treatment resistant OCD with an equal proportion of patients benefiting from right- and left-sided stimulation [110]. In an open-label trial,10 patients with a history of several medication trial failures were treated for 2 weeks with low-frequency TMS. The supplementary motor area (SMA) was chosen as the site of stimulation with the coil placed along the sagittal midline in order to stimulate the SMA bilaterally and simultaneously. The sample had a significant general clinical improvement on the Clinical Global Impression Scale at the end of the first and second week of treatment, and maintained that benefit at 1 month and 3 month follow-up [111]. Chiara Ruffini and colleagues took twenty-three consecutively admitted right-handed inpatients with DSM-IV-TR–diagnosed drug-resistant OCD and gave rTMS to the left OFC parallel (active: n = 16) or perpendicular (sham: n = 7) to the scalp. The patients' OCD symptoms, mood, and anxiety were rated at baseline, at the end of treatment, and once every 2 weeks for 3 months after treatment. Considering changes in Yale-Brown Obsessive Compulsive Scale (YBOCS) scores with 2-way analysis of variance for repeated measures for a total of 8 observations (before rTMS, after treatment, and every 2 weeks for 12 weeks' follow-up), and found significant reduction of YBOCS scores comparing active versus sham treatment for 10 weeks after the end of rTMS ($P < .02$), with loss of significance after 12 weeks ($P < .06$). They concluded Low-frequency rTMS of the left OFC produced significant but time-limited improvement in OCD patients compared to sham treatment [112]. In a Randomized sham-controlled trial of repetitive transcranial magnetic stimulation in treatment-resistant obsessive–compulsive disorder Antonio Mantovani et al found the response rate was 67% (6/9) with active and 22% (2/9) with sham rTMS after 4 wk and patients receiving active rTMS showed on average a 25% reduction in the YBOCS compared to a 12% reduction in those receiving sham. In those who received 8-wk active rTMS, OCD symptoms improved from 28.2 ± 5.8 to 14.5 ± 3.6 [113].

Albert et al(2013) have most recently extensively reviewed available data on treatment resistant OCD and attempted to build a treatment algorithm for those patients who fail to respond to a first SSRI trial [114].

As WHO considers OCD to be fourth major mental illness with significant morbidity we need more effective treatment options. Though a lot of research is available regarding evidence based treatment approaches in treatment refractory OCD, a lot of clinical mysteries are still present and further research need to be done. Most of the research available today is preliminary short duration studies with small sample size and we need long term clinical trials to further validate the hypothesis formed from these preliminary studies.

Serotonin related gene polymorphism is a field which might help in the prediction of response to serotonergic drugs. Denys et al found that response in venlafaxine-treated OCD patients is associated with the S/L genotype of the 5-HTTLPR polymorphism and in paroxetine-treated OCD patients with the G/G genotype of the 5-HT2A polymorphism [115].

Functional imaging techniques for glucose metabolism might also predict treatment response in patients with OCD. Saxena et al suggested that elevated activity (cerebral glucose metabolism) in the right caudate may be a marker of responsiveness to antiobsessional treatment [116].

Metabotropic glutamate receptor anatagonists, are being developed by pharmaceutical industry which might be of help in treatment resistant OCD patients [117, 118, 119].

Some of this exciting research may soon get translated into relevant clinical treatment options for the patients suffering from continued agonizing symptoms of OCD.

Author details

Mushtaq A. Margoob[1], Rajesh Chandel[2], Huda Mushtaq[3] and Dhuha Mushtaq[4]

1 Department of Psychiatry, Institute of Mental Health and Neurosciences-Kashmir, India

2 Department of Psychiatry, Institute of Mental Health and Neurosciences-Kashmir, India

3 Behavioral Genetics and Neurobiology Research Laboratory, Institute of Mental Health and Neurosciences-Kashmir, India

4 Department of Psychiatry, Institute of Mental Health and Neurosciences-Kashmir, India

References

[1] Kaplan & Sadock's Synopsis of Psychiatry: Behavioral Sciences/Clinical Psychiatry, 10th Edition.

[2] WHO's The ICD-10 Classification of Mental and Behavioral Disorders.

[3] Diagnostic and statistical manual of mental disorders, Fifth Edition.

[4] Michael Kellner, Drug treatment of obsessive-compulsive disorder, Dialogues Clin Neurosci. 2010 June; 12(2): 187–19.

[5] Textbook Of Anxiety Disorder/ DJ Stein APA 2nd edition.

[6] J Clin Psychiatry. 1990 Aug;51 Suppl:27-31; discussion 55-8. Neurobiology of obsessive compulsive disorder: a possible role for serotonin. Winslow JT, Insel TR.

[7] J Clin Psychiatry. 2004;65 Suppl 14:11-7. The role of dopamine in obsessive-compulsive disorder: preclinical and clinical evidence. Denys D, Zohar J, Westenberg HG.

[8] Psychiatr Hung. 2007;22(4):248-58.New approach to obsessive-compulsive disorder: dopaminergic theories; Harsányi A, Csigó K, Demeter G, Németh A.

[9] Bronstein Y, Cummings J. In: Lichter D, Cummings J Ed, Frontal-subcortical circuits inpsychiatric and neurological disorders. Guilford, New York. 2001; 59-91.

[10] Rosenberg DR, MacMillan SN, Moore GJ. Brain anatomy and chemistry may predict treatment response in paediatric obsessive-compulsive disorder. Int J Neuropsychopharmacol 2001; 4(2): 179 190.

[11] Moore GJ, MacMaster FP, Stewart C, Rosenberg DR. Case study: caudate glutamatergic changes with paroxetine therapy for pediatric obsessive-compulsive disorder. J Am Acad Child Adolesc Psychiatry 1998: 37(6): 663-67.

[12] Rosenberg DR, MacMaster FP, Keshavan MS, Fitzgerald KD, Stewart CM, Moore GJ. Decrease in caudate glutamatergic concentrations in pediatric obsessive-compulsive disorder patients taking paroxetine. J Am Acad Child Adolesc Psychiatry 2000; 39(9): 1096-103.

[13] Whiteside SP, Port JD, Deacon BJ, Abramowitz JS. A magnetic resonance spectroscopy investigation of obsessive-compulsive disorder and anxiety. Psychiatry Res 2006; 146(2): 137-47.

[14] Dickel DE, Veenstra-VanderWeele J, Cox NJ, et al. Association testing of the positional and functional candidate gene SLC1A1/EAAC1 in early-onset obsessive-compulsive disorder. Arch Gen Psychiatry 2006; 63(7): 778-85.

[15] Arnold PD, Sicard T, Burroughs E, Richter MA, Kennedy JL. Glutamate transporter gene SLC1A1 associated with obsessivecompulsive disorder. Arch Gen Psychiatry 2006; 63(7): 769-776.

[16] Leckman JF, Kim Y. A Primary candidate gene for obsessivecompulsive disorder. Arch Gen Psychiatry 2006; 63(7): 717-20.

[17] Keuler DJ, Altemus M, Michelson D, Greenberg B, Murphy DL. Behavioral effects of naloxone infusion in obsessive-compulsive disorder. Biol Psychiatry 1996; 40: 154-156.

[18] Altemus M, Pigott T, Kalogeras KT, Demitrack M, Dubbert B. Abnormalities in the regulation of vasopressin and corticotrophin releasing factor secretion in obsessive-compulsive disorder. Arch Gen Psychiatry 1992; 49: 9-20.

[19] Leckman JF, Goodman WK, North WG, et al. The role of central oxytocin in obsessive compulsive disorder and related normal behavior. Psychoneuroendocrinology 1994; 19(8): 723-49.

[20] Bhattacharyya S, Chakraborty K; Recent Pat CNS Drug Discov. 2007 Jan;2(1):47-55. Glutamatergic dysfunction--newer targets for anti-obsessional drugs.

[21] Rauch SL,JenikeMA(1994)Management of treatment resistant obsessive compulsive disorder: concepts and strategies. In Hollander E, Zohar J,Marazziti D, Olivieri B(eds) current insights in obsessive compulsive disorder,John Wiley and Sons Ltd, Chichester.

[22] Stein DJ, Ipser JC, Baldwin DS, Bandelow B. Treatment of obsessive-compulsive disorder.. CNS Spectr. 2007;12(suppl 3):28–35.

[23] DS, Anderson IM, Nutt DJ, et al. Evidence-based guidelines for the pharmacological treatment of anxiety disorders: recommendations from the British Association for Psychopharmacology.. J Psychopharmacol. 2005;19:567–596.

[24] American Psychiatric Association: Practice guidelines for the treatment of patients with obsessive-compulsive disorder.. Am J Psychiatry. 2007;164 (suppl 1):1–56.

[25] W B, Zohar J, Hollander E, et al. World federation of societies of biological psychiatry (WFSBP) guidelines for the pharmacological treatment of anxiety, obsessive-compulsive and post-traumatic stress disorders first revision.. W J Biol Psychiatry. 2008;9:248–312.

[26] NA, Pampaloni I, Pallanti S, Ipser J, Stein DJ. Sustained response versus relapse: the pharmacotherapeutic goal for obsessive-compulsive disorder.. Int Clin Psychopharmacol. 2007;22:313–322.

[27] NICE clinical guideline 31 Obsessive-compulsive disorder: core interventions in the treatment of obsessive-compulsive disorder and body dysmorphic disorder.. 2005 Available at: www. Nice.org.uk.

[28] GM, Altmann DG, Rajagopal S, Oakley-Browne M. Selective serotonin re-uptake inhibitors (SSRIs) versus placebo for obsessive compulsive disorder (OCD).. Cochrane Database Syst Rev. 2008;1.

[29] McDougle CJ, Goodman WK, Leckman JF, Price LH. The psychopharmacology of obsessive-compulsive disorder implications for treatment and pathogenesis.. Psychiatr Clin North Am. 1993;16:749–766.

[30] Hollander E, Kwon JH, Stein DJ, Broatch J, Rowland CT, Himelein CA. Obsessive-compulsive and spectrum disorders: overview and quality of life issues. J Clin Psychol 57:3– 6, 1996.

[31] Goodman WK, Price LH, Rasmussen SA, Mazure C, Fleischmann RL, Hill CL, Heninger GR, Charney DS. The Yale-Brown obsessive compulsive scale. I. Development, use, and reliability. Arch Gen Psych 46:1106 –1111, 1989.

[32] Goodman WK, Price LH, Rasmussen SA, Mazure C, Delgado P, Heninger GR, Charney DS. The Yale-Brown obsessive compulsive scale. II. Validity. Arch Gen Psych 46:1012–1016, 1989.

[33] Ninan pT, Koran LM, Kiev A, et al. High-dose sertraline strategy for nonresponders to acute treatment for obsessive-compulsive disorder: a multicenter double-blind trial.. J Clin Psychiatry. 2006;67:15–22.

[34] Rabinowitz I Baruch Y, Barak Y. High-dose escitalopram for the treatment of obsessive-compulsive disorder.. Int Clin Psychopharmacol. 2008;32:49–53.

[35] Dougherty DD, Jameson M, Deckersbach T, Loh R, Thompson-Hollands J, Jenike M, Keuthen NJ. Open-label study of high (30 mg) and moderate (20 mg) dose escitalopram for the treatment of obsessive-compulsive disorder. Int Clin Psychopharmacol. 2009 Nov;24(6):306-11.

[36] Insel TR, Murphy DL, Cohen RM, Alterman I, Kilts C, Linnoila M. Obsessive-compulsive disorder—a double-blind trial of clomipramine and clorgyline. Arch Gen Psychol 40:605– 612, 1983.

[37] Geller DA, Biederman J, Stewart SE, Mullin B, Martin A, Spencer T, Faraone SV. Which SSRI? A meta-analysis of pharamacotherapy trials in pediatric obsessive-compulsive disorder. Am J Psychol 160:1919 –1928, 2003.

[38] Fineberg NA, Gale TM. Evidence-based pharmacotherapy of obsessive- compulsive disorder. Int J Neuropsychopharm 8:107–129,2005.

[39] Hollander E, Friedberg J, Wasserman S, Allen A, Birnbaum M, Koran LM. Venlafaxine in treatment-resistant obsessive-compulsive disorder. J Clin Psychol 64:972, 2003.

[40] Albert U, Aguglia E, Maina G, Bogetto F. Venlafaxine versus clomipramine in the treatment of obsessive-compulsive disorder: a preliminary single-blind, 12-week, controlled study. J Clin Psychol 63:1004 –1009, 2002.

[41] Pallanti S, Quercioli L, Bruscoli M. Response acceleration with mirtazapine augmentation of citalopram in obsessive compulsive disorder patients without comorbid depression: a pilot study. J Clin Psychol 65:1394 –1399, 2004.

[42] Schule C, Laakmann G. Mirtazapine plus citalopram has short term but not longer term benefits over citalopram alone for the symptoms of obsessive compulsive disorder. Evid Based Ment Health 8:42, 2005.

[43] Koran LM, Gamel NN, Choung HW, Smith EH, Aboujaoude EN. Mirtazapine for obsessive-compulsive disorder: an open trial followed by double-blind discontinuation. J Clin Psychol 66:515– 520, 2005.

[44] Koran LM, Faravelli CF, Pallanti S. Intravenous clomipramine in obsessive-compulsive disorder. J Clin Psychopharm 14:216 –218, 1994.

[45] Fallon BA, Liebowitz MR, Campeas R, Schneier FR, Marshall R, Davies S, Goetz D, Klein DF. Intravenous clomipramine for obsessive-compulsive disorder refractory to oral clomipramine: a placebo-controlled study. Arch Gen Psychol 55:918 –924, 1998.

[46] Pallanti S, Quercioli L, Koran LM. Citalopram intravenous infusion in resistant obsessive-compulsive disorder. J Clin Psychol 63:796–801, 2002.

[47] Stahl SM. Finding what you are not looking for: strategies for developing novel treatments in psychiatry. NeuroRx 3:000–000, 2006.

[48] Leckman JF, Walker DE, Goodman WK, Pauls DL, Cohen DJ. "Just right" perceptions associated with compulsive behavior in Tourette's syndrome. Am J Psychol 151:675– 680, 1994.

[49] Graybiel AM, Rauch SL. Toward a neurobiology of obsessivecompulsive disorder. Neuron 28:343–347, 2000.

[50] Leckman JF, Riddle MA. Tourette's syndrome: when habit-forming systems form habits of their own? Neuron 28:349 –354, 2000.

[51] Scahill L, Leckman JF, Schultz RT, Katsovich L, Peterson BS. A placebo-controlled trial of risperidone in Tourette syndrome. Neurology 60:1130 –1135, 2003.

[52] Dion Y, Annable L, Sandor P, Chouinard G. Risperidone in the treatment of Tourette syndrome: a double-blind, placebo-controlled trial. J Clin Psychopharm 22:31–39, 2002.

[53] Savas HA, Yumru M, Ozen ME. Quetiapine and ziprasidone as adjuncts in treatment-resistant obsessive-compulsive disorder: a retrospective comparative study.. Clin Drug Investig. 2008;28:439–442.

[54] Maina G, Pessina E, Albert, Bogetto F. 8-week, single-blind, randomized trial comparing risperidone versus olanzapine augmentation of serotonin reuptake inhibitors in treatment-resistant obsessive-compulsive disorder.. Eur Neuropsychopharmacol. 2008;18:364–372.

[55] Pessina E, Albert U, Bogetto F, Maina G. Aripiprazole augmentation of serotonin reuptake inhibitors in treatment-resistant obsessive-compulsive disorder: a 12-week open-label preliminary study.. Int Clin Psychopharmacol. 2009;24:265–269.

[56] Connor KM, Payne VM, Gadde KM, Zhang W, Davidson JR. The use of aripiprazole in obsessive-compulsive disorder: preliminary observations in 8 patients.. J Clin Psychiatry. 2005;66:49–51.

[57] Otsuka T, Togo T, Sugiyama N, et al. Perspirone augmentation of paroxetine in treatment of refractory obsessive-compulsive disorder with depression.. Prog Neuropsychopharmacol Biol Psychiatry. 2007;32:564–568.

[58] Koran LM, Aboujaoude E, Bullock KD, Franz B, Gamel N, Elliot M. Double-blind treatment with oral morphine in treatment-resistant obsessive-compulsive disorder. J Clin Psychol 66:353– 359, 2005.

[59] Shapira NA, Keck PE Jr, Goldsmith TD, McConville BJ, Eis M, McElroy SL. Open-label pilot study of tramadol hydrochloride in treatment-refractory obsessive-compulsive disorder. Depress Anxiety 6:170 –173, 1997.

[60] Goldsmith TB, Shapira NA, Keck PE Jr. Rapid remission of OCD with tramadol hydrochloride. Am J Psychol 156:660–661, 1999.

[61] Hollander E, Bienstock CA, Koran LM, Pallanti S, Marazziti D, Rasmussen SA, Ravizza L, Benkelfat C, Saxena S, Greenberg BD, Sasson Y, Zohar J. Refractory obsessive-

compulsive disorder: state-of-the-art treatment. J Clin Psychol 63 [Suppl 6]:20–29, 2002.

[62] Marazziti D, Golia F, Consoli G, et al. Effectiveness of long-term augmentation with citalopram to clomipramine in treatment-resistant OCD patients.. CNS Spectr. 2008;13:971–976.

[63] Aventis: Rilutek. In: Physician's desk reference. Montvale, NJ: Thompson Healthcare, 2004.

[64] Urbani A, Belluzzi O. Riluzole inhibits the persistent sodium current in mammalian CNS neurons. Eur J Neurosci 12:3567– 3574, 2000.

[65] Wang SJ, Wang KY, Wang WC. Mechanisms underlying the riluzole inhibiton of glutamate release from rat cerebral cortex nerve terminals (synaptosomes). Neuroscience 125:191–201, 2004.

[66] Xu L, Enyeart JA, Enyeart JJ. Neuroprotective agent riluzole dramatically slows in activation of Kv1.4 potassium channels by a voltage-dependent oxidative mechanism. J Pharm Exp Ther 299:227–237, 2001.

[67] Frizzo ME, Dall'Onder LP, Dalcin KB, Souza DO. Riluzole enhances glutamate uptake in rat astrocyte cultures. Cell Mol Neurobiol 24:123–128, 2004.

[68] Coric V, Taskiran S, Pittenger C, Wasylink S, Mathalon DH, Valentine G, Saksa J, Wu YT, Gueorguieva R, Sanacora G, Malison RT, Krystal JH. Riluzole augmentation in treatmentresistant obsessive-compulsive disorder: an open-label trial. Biol Psychol 424–428, 2005.

[69] Coric V, Milanovic S, Wasylink S, Patel P, Malison R, Krystal JH. Beneficial effects of the antiglutamatergic agent riluzole in a patient diagnosed with obsessive-compulsive disorder and major depressive disorder. Psychopharm (Berl) 167:219 –220, 2003.

[70] Moran MM, McFarland K, Melendez RI, Kalivas PW, Seamans JK. Cystine/glutamate exchange regulates metabolic glutamate receptor presynaptic inhibition of excitatory transmission and vulnerability to cocaine seeking. J Neurosci 25:6389–6393, 2005.

[71] Lafleur DL, Pittenger C, Kelmendi B, Gardner T, Wasylink S, Malison RT, Sanacora G, Krystal JH, Coric V. N-acetylcysteine augmentation in serotonin reuptake inhibitor refractory obsessive-compulsive disorder. Psychopharm (Berl) 184:254–256, 2006.

[72] Pittinger C, Krystal JH, Coric V. Initial evidence of the beneficial effects of glutamate modulating agents in the treatment of self injurious behavior associated with borderline personality disorder. J Clin Psychol 66:1492–1493, 2005.

[73] Aboujaoude E, Barry JJ, Gamel N. Memantine augmentation in treatment-resistant obsessive-compulsive disorder an open label trial.. J Clin Psychopharmacol. 2009;29:51–55.

[74] Feusner JD, Kerwin L, Saxena S, Bystritsky A, Differential efficacy of memantine for obsessive-compulsive disorder vs. generalized anxiety disorder: an open-label trial. Psychopharmacol Bull. 2009;42(1):81-93.

[75] Stewart, S. Evelyn Jenike, Eric A. Hezel, Dianne M.; Stack, Denise Egan; Dodman, Nicholas H.; Shuster, Louis; Jenike, Michael A. A Single-Blinded Case-Control Study of Memantine in Severe Obsessive-Compulsive Disorder,Journal of Clinical Psychopharmacology: February 2010 - Volume 30 - Issue 1 - pp 34-39.

[76] Greenberg WM, Benedict MM, Doerfer J, et al. Adjunctive glycine in the treatment of obsessive-compulsive disorder in adults.. J Psychiatr Res. 2009;43:664–670.

[77] Egashira N, Okuno R, Harada S, et al. Effects of glutamate-related drugs on marble-burying behavior in mice: implications for obsessive-compulsive disorder.. Eur J Pharmacol. 2008;586:164–170.

[78] Hollander E, Dell'Osso B. Topiramate plus paroxetine in treatment-resistant obsessive-compulsive disorder.. Int Clin Psychopharmacol. 2006;21:189–191.

[79] van Ameringen M, Mancini C, Patterson B, Bennett M. Topiramate augmentation in treatment-resistant obsessive-compulsive disorder: a retrospective, open-label case series.. Depress Anxiety. 2006;23:1–5.

[80] ClinicalTrials.gov Identifiers: NCT00211744 and NCT00182520.

[81] oulis P,Masdrakis VG, Karapoulios E, et al. Pregabalin augmentation to sertraline-risperidone combination in the treatment of obsessive-compulsive disorder. Prim Care Companion. J Clin Psychiatry. 2008;10:249.

[82] Roebel M, Nenadic I, Sauer H. Pregablin as a successful pharmacotherapeutic option in OCD - a case report.. Nervenarzt. 2009;80(suppl 2):P–043-007.

[83] ClinicalTrials.gov Identifier: NCT00994786.

[84] Kushner MG, Kim SW, Donahue C, et al. D-cycloserine augmented exposure therapy for obsessive-compulsive disorder.. Biol Psychiatry. 2007;15:835–838.

[85] Wilhelm S, Buhlmann U, Tolin DF, et al. Augmentation of behavior therapy with D-cycloserin for obsessive-compulsive disorder.. Am J Psychiatry. 2008;165:335–341

[86] Eric A. Storch, Tanya K. Murphy, Wayne K. Goodman, Gary R. Geffken, Adam B. Lewin, Aude Henin, Jamie A. Micco, Susan Sprich, Sabine Wilhelm, Michael Bengtson, Daniel A. Geller, A Preliminary Study of D-Cycloserine Augmentation of Cognitive-Behavioral Therapy in Pediatric Obsessive-Compulsive Disorder; Biological Psychiatry Volume 68, Issue 11, 1 December 2010, Pages 1073–1076.

[87] Fujiwara M, Egashira N: New perspectives in the studies on en- docannabinoid and cannabis: abnormal behaviors associate with CB1-receptor and development of therapeutic applica-tion. J Pharmacol Sci 2004; 96:363–366.

[88] Pacher P, Bátkai S, Kunos G: The endocannabinoid system as an emerging target of pharmacotherapy. Pharmacol Rev 2006; 58:389–462.

[89] Improvement in Refractory Obsessive Compulsive Disorder With Dronabinol ;Am J Psychiatry 2008;165:536-537. 10.1176/appi.ajp.2007.07061016.

[90] Simpson HB, Foa EB, Liebowitz MR, Ledley DR, Huppert JD, Cahill S, Vermes D, Schmidt AB, Hembree E, Franklin M, Campeas R, Hahn CG, Petkova E.; A randomized, controlled trial of cognitive-behavioral therapy for augmenting pharmacotherapy in obsessive-compulsive disorder.; Am J Psychiatry. 2008 May;165(5):621-30. doi: 10.1176/appi.ajp.2007.07091440. Epub 2008 Mar 3.

[91] Tolin DF, Maltby N, Diefenbach GJ, Hannan SE, Worhunsky P. Cognitivebehavioral therapy for medication nonresponders with obsessive-compulsive disorder: a wait-list-controlled open trial.. J Clin Psychiatry. 2004;65:922–931.

[92] Eddy KT, Dutra L, Bradly R Westen D. A multidimensional meta-analysis of psychotherapy and pharmacotherapy for obsessive-compulsive disorder.. Clin Psychol Rev. 2004;24:1011–1030.

[93] Christopher Pittenger, John H. Krystal, and Vladimir Coric, Glutamate-Modulating Drugs as Novel Pharmacotherapeutic Agents in the Treatment of Obsessive-Compulsive Disorder: The Journal of the American Society for Experimental NeuroTherapeutics; Vol. 3, 69–81, January 2006.

[94] Nuttin B, Cosyns P, Demeulemeester H, Gybels J, Meyerson B (1999). Electrical stimulation in anterior limbs of internal capsules in patients with obsessive–compulsive disorder. Lancet 354.

[95] Nuttin BJ, Gabriels LA, Cosyns PR, Meyerson BA, Andreewitch S, Sunaert SGet al (2003b). Long-term electrical capsular stimulation in patients with obsessive–compulsive disorder. Neurosurgery 52: 1263–1272.

[96] Gabriels L, Cosyns P, Nuttin B, Demeulemeester H, Gybels J (2003). Deep brain stimulation for treatment-refractory obsessive–compulsive disorder: psychopathological and neuropsychological outcome in three cases. Acta Psychiatr Scand 107: 275–282.

[97] Abelson JL, Curtis GC, Sagher O, Albucher RC, Harrigan M, Taylor SF et al(2005). Deep brain stimulation for refractory obsessive–compulsive disorder.Biol Psychiatry 57: 510–516.

[98] Anderson D, Ahmed A (2003). Treatment of patients with intractable obsessive–compulsive disorder with anterior capsular stimulation. Case report. J Neurosurg 98: 1104–1108.

[99] Aouizerate B, Cuny E, Martin-Guehl C, Guehl D, Amieva H, Benazzouz A et al(2004). Deep brain stimulation of the ventral caudate nucleus in the treatment of obsessive–compulsive disorder and major depression. Case report. J Neurosurg 101: 682–686.

[100] Sturm V, Lenartz D, Koulousakis A, Treuer H, Herholz K, Klein JC et al(2003). The nucleus accumbens: a target for deep brain stimulation in obsessive–compulsive- and anxiety-disorders. J Chem Neuroanat 26: 293–299.

[101] Aouizerate B, Martin-Guehl C, Cuny E, Guehl D, Amieva H, Benazzouz A et al(2005). Deep brain stimulation for OCD and major depression. Am J Psychiatry 162: 2192.

[102] Benjamin D Greenberg, Donald A Malone, Gerhard M Friehs,et al; Three-Year Outcomes in Deep Brain Stimulation for Highly Resistant Obsessive–Compulsive Disorder; Neuropsychopharmacology (2006) 31, 2384–2393. doi:10.1038/sj.npp.1301165; published online 19 July 2006.

[103] Maletzky B, McFarland B, Burt A. Refractory obsessive compulsive disorder and ECT. Convuls Ther. 1994;10:34-42.

[104] Thomas SG, Kellner CH. Remission of major depression and obsessiveCompulsive disorder after a single unilateral ECT. J ECT. 2003;19:50-51.

[105] Lavin MR, Halligan P. ECT for obsessive-compulsive disorder and schizophrenia. Am J Psychiatry. 1996;153:1652-1653.

[106] Casey DA, Davis MH. Obsessive-compulsive disorder responsive to electroconvulsive therapy in an elderly woman. South Med J. 1994;87:862-864

[107] Mellman LA, Gorman JM. Successful treatment of obsessive-compulsive disorder with ECT. Am J Psychiatry. 1984;141:596-597.

[108] Strassnig M, Riedel M, Muller N. Electroconvulsive therapy in a patient with Tourette's syndrome and co-morbid obsessive compulsive disorder. World J Biol Psychiatry. 2004;5:164-166.

[109] Greenberg BD, George MS, Martin JD, et al. Effect of prefrontal repetitive transcranial magnetic stimulation in obsessive-compulsive disorder: a preliminary study. Am J Psychiatry. 1997;154:867-869.

[110] Sachdev PS, McBride R, Loo CK, Mitchell PB, Malhi GS, Croker VM. Right versus left prefrontal transcranial magnetic stimulation for obsessive-compulsive disorder: a preliminary investigation. J Clin Psychiatry. 2001;62:981-984.

[111] Mantovani A, Lisanby SH, Pieraccini F, et al. Repetitive transcranial magnetic stimulation (rTMS) in the treatment of obsessive-compulsive disorder (OCD) and Tourette's syndrome (TS). Int J Neuropsychoph. 2005 Jun 28;1-6.

[112] Chiara Ruffini, Marco Locatelli,, Adelio Lucca, Augmentation Effect of Repetitive Transcranial Magnetic Stimulation Over the Orbitofrontal Cortex in Drug-Resistant Obsessive-Compulsive Disorder Patients: A Controlled Investigation Prim Care Companion J Clin Psychiatry. 2009; 11(5): 226–230.

[113] Antonio Mantovani, Helen B. Simpson, Brian A. Fallon, Simone Rossi and Sarah H. Lisanby Randomized sham-controlled trial of repetitive transcranial magnetic stimu-

http://dx.doi.org/10.5772/57232

lation in treatment-resistant obsessive–compulsive disorder The International Journal of Neuropsychopharmacology / Volume 13 / Issue 02 /March 2010, pp 217-227.

[114] Umberto Albert, Andrea Aguglia, Stefano Bramante, Filippo Bogetto, Giuseppe Maina: TREATMENT RESISTANT OBSESSIVE COMPULSIVE DISORDER(OCD): CURRENT KNOWLEDGE AND OPEN QUESTIONS:Clinical Neuropsychiatry (2013) 10,1,19-30.

[115] Denys D, Van Nieuwerburgh F, Deforce D, Westenberg HG. Prediction of response to paroxetine and venlafaxine by serotonin-related genes in obsessive-compulsive disorder in a randomized, double-blind trial.J Clin Psychiatry. 2007 May;68(5):747-53.

[116] Saxena S, Brody AL, Ho ML, Zohrabi N, Maidment KM, Baxter LR Jr. Differential brain metabolic predictors of response to paroxetine in obsessive-compulsive disorder versus major depression. Am J Psychiatry. 2003 Mar;160(3):522-32.

[117] Edwards, I., Isaac, M., Johansson, M., Malmberg, J., Mindis, A.,Staaf, K., Slassi, A., Wensbo, D.: WO05080379A1 (2005).

[118] Bryan C, Isaac M, Stefanac T: WO05080363A1 (2005).

[119] Johansson, M., Wensbo, D., Mindis, A., Staaf, K., Edwards, L., Isaac, M., Stefanac, T., Slassi, A., McLeod, D.: WO05080397A2(2005).

Section 3

Obsessive-Compulsive Disorder and Other Neuropsychiatric Disorders

Chapter 8

Neurobiological and Personality Risk Factors for Development of Obsessive-Compulsive Disorder in Patients with Epilepsy

Vladimir V. Kalinin, Anna A. Zemlyanaya,
Elena V. Zheleznova and Lyudmila V. Sokolova

Additional information is available at the end of the chapter

http://dx.doi.org/10.5772/58230

1. Introduction

Psychiatric co-morbidities in forms of affective and anxiety disorders are believed to be the most frequent complications of epilepsy. It concerns mostly the patients with localization related forms, and temporal lobe epilepsy [TLE], in particular. Perini et al. have shown that patients with TLE have higher rates of affective and personality disorders compared to persons with juvenile myoclonic epilepsy [1]. Despite their frequent prevalence that achieves about 8%-50% among patients with epilepsy [2-6] the data on the risk factors for the development of psychiatric co-morbidities in TLE remain obscure and rather controversial. That also concerns obsessive-compulsive disorder (OCD), which is regarded as one of co-morbid psychopathological complications in TLE [7].

According to Bear and Fedio data [8] focus laterality in TLE determines the kind of interictal psychopathology, and in patients with right-sided TLE such symptoms as sadness, elation and obsessional thinking can origin, while in the left-sided TLE paranoia, anger, dependence and negative self-image may appear.

Nevertheless, these results have been lately criticized since some subsequent studies couldn't replicate these data [9,10]. It implies that data on foci lateralization themselves couldn't explain the origin of some psychopathological signs and obsessional thinking in particular in all cases of TLE. Obviously some other neurobiological factors and personality traits should also be taken into account in order to predict the probable OCD development in TLE. Moreover, whether the seizure semiotics has influence upon its development or not remains unknown although in 19-th century have been reported that obsessions were frequently accompanied by a family history of epilepsy [11, 12].

Among neurobiological factors the data on motor lateralization can be important for prediction of psychopathological comorbidity i.e. OCD. Along with neurobiological signs the personality structure variables also may be important for this reason.

Thus, focus lateralization in TLE is thought not to be enough to predict the probability of psychopathology development in patients with epilepsy, and other factors, such as motor lateralization and premorbid personality features also should be taken into account. In one of our previous study we have shown that Alexithymia exerts maximal effect on psychopathological variables, i.e. Depression in SCL-90, and maximal values of SCL-90 constructs were observed in cases with combination Alexithymia and left-handedness and Alexithymia and right-sided seizure focus [13]. Here should be stressed that Alexithymia itself represents only a part of complicated psychopathological structure of so-called premorbid personality, and other personality factors may be also important for the development of psychopathological disorders in patients with TLE. Any possible relationships between personality, neurobiological variables and OCD in epilepsy have not been properly studied.

Premorbid personality features to date have been fairly studied in terms of risk for psychiatric pathology in numerous studies. It concerned mainly the role of premorbid personality structure in affective disorders, schizophrenia, and organic disorders, although the strictly definite and universal findings have not been obtained yet. Thus, Tellenbach introduced concept of "Typus melancholicus" (TM) which is still regarded as premorbid personality features constellation for unipolar (recurrent) depression development [14-16]. Such approach has been mainly used in Germany and Japan and obtained in this context data are rather controversial, and complicated. Thus, if some authors agree in unison on the significant role of TM in depression development [14-17] other authors are careful and could find, that only between 30 and 70 % of depressed patients have a typical TM structure [18, 19].

Thus, Furukawa et al.[19] used a special scales for assessment of TM features in 140 psychiatric patients and couldn't find any statistically significant discrepancies with normal control subjects except that among the recurrent depression patients the TM score was lower than in control group. Unfortunately the similar study has not been performed in epilepsy patients yet, and whether the premorbid personality has influence on the development of psychopathology remains unknown.

To date the several personality traits are thought to be significant for the development of psychiatric disorders at all, and some tests have been elaborated in this context to detect such personality predisposition to depression in particular. One of such questionnaire widely known as Munich Personality Test (MPT) has been developed by von Zerssen et al using factor analysis in 1988 [20]. Detailed description of MPT is given in Materials and Methods section.

Although such multidimensional approach has been widely used in psychiatry, data on its use in patients with epilepsy are practically absent. Since such neuropsychological variables as focus laterality (FL) and motor lateralization (ML) are important for development of psychiatric disorders in patients with TLE [13], suggestion could be made that premorbid personality

http://dx.doi.org/10.5772/58230

features in patients with TLE might be significant for the development of comorbid psychopathology, and OCD in particular, if take into account their interaction between each other.

The principal aim of the current study was to find any possible relationships between personality features and OCD development in patients with different neurobiological patterns i.e. in persons with different combination of FL and ML. The corner stone of the current study was hypothesis, that the different combinations of FL and ML are thought to be a neurobiological basis for relationship between premorbid personality features, on the one hand, and obsessive-compulsive symptoms, on the other hand in patients with TLE.

2. Material and methods

The study has been carried out on 103 patients with epilepsy. Among all studied patients were 33 men and 70 women. The diagnosis of symptomatic epilepsy was set in 40 patients, the diagnosis of cryptogenic form –in 54 patients, and the diagnosis of idiopathic temporal lobe epilepsy-in 9 patients. The focus laterality was detected strictly by visual EEG-method, and data on ictal semiotics have not been taken into account. The left-sided foci were detected in 48 patients (12 men and 36 women), the right-sided foci – in 55 patients (22 men and 33 women).

All patients were evaluated by psychiatrists in order to set a psychiatric diagnosis. ICD-10 criteria were used for these purposes. In line with these criteria the diagnosis of OCD was set in 19 patients. For the assessment of severity of OCD symptoms the correspond construct of SCL-90 scale was used and filled in by physicians. The mean value of obsessive-compulsive construct in OCD patients reached 20,16+-4,13, while in patients without OCD – only 6,31+-3, 72 points.

For assessment of the premorbid personality features the MPT has been used [20]. The MPT represents a self-rating questionnaire, and includes 51 questions depicting the different personality traits. The patients have filled in all rating scales themselves, and after that the obtained raw data have been transformed into six constructs in line with specific structure of scales. These constructs include Extraversion, Neuroticism, Rigidity, Frustration Tolerance, Tendencies to Isolation and Esoteric Tendency. The last two constructs form Schizoidia scale [20]. The other two control scales of MPT (Orientation towards Social Norms and Motivation) were not included in the final analysis.

Extraversion and Neuroticism constructs are derived from Eysenck and Eysenck concepts [21]. Rigidity is quite similar to construct of Typus melancholicus proposed by Tellenbach [14-17], while Tendency to Isolation and Esoteric Tendency are based on Kretschmer's classical study on relationships between constitution and personality [22]. Frustration tolerance refers to resiliency or stress coping strategy.

Along with MPT the Toronto Alexithymia Scale (TAS-26) [23, 24] was explored for assessment of alexithymia. This scale consists of 26 items, and each item can be scored in points from 1 to 5. The global alexithymia score in TAS-26 may be expressed from 26 to 130 points. All patients whose global TAS-26 score exceeds 74 points were regarded as alexithymic persons. The mean average of TAS-26 score in nonalexithymic group (N=78) was 58,5+-11,5, and in alexithymic group (N=22) was 80,0+-4,8 points.

Assessment of psychopathological status of patients has been performed by use Symptom Check List-90 (SCL-90). This questionnaire represents a self-rated scale that has 9 psychiatric symptom groups, consisting of 90 items with a range of five degree severity (0,1,2,3,4). The evaluated psychiatric constructs include somatization, obsessive-compulsive symptoms, interpersonal sensitivity, depression, anxiety, hostility, phobic anxiety, paranoid ideations, and psychoticism [25, 26].

For the assessment of handedness Annet's scale was used [27]. Persons whose global score on that scale was lower than-5 points were regarded as left-handers, whilst persons with global score exceeded+5 points-as right-handers. Among all studied patients 73 persons were considered as right-handers (Mean+-Std. Dev.:+19,2+-6,0) in Annet's score and 31 persons as left-handers (Mean+-Std. Dev:-11,2+-10,9).

For analysis purposes al patients were divided into groups in relation to combination of focus lateralization and motor asymmetry (handedness). In line with this rule the next four groups were defined: Left focus/Left-handedness (13 patients), Left focus/Right-handedness (33 patients), Right focus/Left-handedness (17 patients) and Right focus/Right-handedness (36 patients).

3. Statistics

All data were statistically processed by Statistica program (9-th version) on personal computer. On the first step the possible relationship between focus laterality and OCD development was assessed. In addition the mean values of seizure frequency of different semiotics in patients with OCD and in persons without OCD have been calculated. On the next step the mean values of premorbid personality traits, including MPT-constructs and TAS-26 have been calculated, while on the third stage the multiple regression analysis has been performed for OCD construct (SCL-90). Here OCD construct was dependent variable, while the premorbid personality features (TAS-26 $\chi$and MPT constructs) were used as independent variables. On the final stage the values of variances obtained in result of regression analysis were compared between different groups of patients, and the maximal variances were regarded as more robust criteria for prediction of OCD [28].

4. Results

The main results are listed in the next several tables. In Table 1 the mean values of seizure frequency are presented. As can be seen no statistically significant discrepancies between groups with and without OCD are obtained. In other words, the seizure semiotics doesn't determine the OCD development. Noteworthy, the statistically significant relationship between focus laterality and OCD also has not been revealed (χ2=2,25, p=0,13), that implies the lack of influence of focus laterality upon OCD development in patients with epilepsy.

http://dx.doi.org/10.5772/58230

Type of seizure	Patients with OCD (n=19)	Patients without OCD (n=84)	Discrepancy
Primary GTCS	0,05+-0,23	0, 27+-0,76	n.s.
Simple partial seizures	25, 68+-69,48	49,23+-136,3	n.s.
Sensory simple partial seizures	5,21+-10,11	37,58+-120,67	n.s.
Motor simple partial seizures	20,68+-70,3	14, 10+-74,63	n.s.
Complex partial seizures	31,47+-71,44	21,36+-57,71	n.s.
Secondary GTCS	6,89+-13,69	7,20+-12,57	n.s.

Notes: GTCS – Generalized tonic-clonic seizures

Table 1. Mean frequency (per year) of different semiotics seizures in patients with OCD and without OCD in epilepsy

The premorbid personality structure data, quite the contrary, have revealed their role in the origin of OCD, as can be seen from Table 2. As can be seen from table the OCD patients compared with patients without OCD symptoms as a whole, were characterized by less degree of Extraversion and Frustration Tolerance than patients in whom OCD symptoms didn't develop. On the other hand, OCD patients in premorbid personality structure had higher values of Neuroticism, Esoteric Tendencies and Schizoidia in comparison with patients without OCD. The values of Alexithymia, Rigidity and Tendencies to Isolation couldn't discriminate two compared groups of patients.

Personality construct	Patients with OCD (n=19)	Patients without OCD (n=84)	Discrepancy
Extraversion (MPT)	11,74+-5,61	16,38+-5,64	p=0,0016
Neuroticism (MPT)	18,26+-3,69	12,37+-5,44	p=0,000012
Frustration Tolerance (MPT)	5,58+-2,85	8,19+-3,98	p=0,0081
Rigidity (MPT)	12,53+-3,91	13,00+-4,56	n.s.
Tendency to Isolation (MPT)	5,37+-2,65	4,67+-2,50	n.s.
Esoteric Tendencies (MPT)	3,37+-2,29	2,27+-2,08	p=0,044
Schizoidia (MPT)	8,89+-3,59	6,93+-3,72	p=0,039
Alexithymia (TAS-26)	68,94+-13,69	62,35+-13,66	n.s.

Table 2. Mean values of different Munich Personality Test constructs and alexithymia in patients with OCD and without OCD in epilepsy

Based on these obtained data preliminary conclusion can be made that interaction of multifarious personality factors seems to be significant for the OCD origin. In order to reveal such interaction the multiple regression analysis (forward model) has been performed in different groups of patients formed on the basis of different combinations motor lateralization and focus laterality. In other words, this analysis has been performed for prediction of OCD level based on premorbid personality variables.

Main results of that analysis are listed in the Table 3.

Group	Alexithymia	Extraversion	Neuroticism	Frustration tolerance	Rigidity	Tend.to isolation	Esoteric tendency	Schizoidia	R2
All persons (N=99)	-	-0,24	0,48	-	-	-	-	-	0,49
RH (n=69)	-	-0,25	0,55	-	-	-0,25	-	-	0,56
LH (n=30)	0,51	-	-	-	-	-	-	-	0,50
RF (n=53)	-	-0,23	0,62	-	-	-0,35	-	-	0,69
LF (n=46)	-	-	-	-	-	-	-	-	0,35
RHRF (n=36)	-	-0,35	0,67	-	-	-0,27	-	-	0,74
RHLF (n=33)	-0,41	-	0,41	-0,37	-	-	-	-	0,51
LHRF (n=17)	0,53	-	0,38	-	-	-	-	-	0,60
LHLF (n=13)	0,145	-0,64	-	-	0,168	0,983	0,79	-0,92	0,78

Notes: RH – right-handers; LH-lefthanders; RF-right focus patients; LF – left focus patients; RHRF-right-handers with right focus; RHLF – right-handers with left focus; LHRF-lefthanders with right focus; LHLF-lefthanders with left focus; R2 – explained variance.

Table 3. Multiple forward stepwise regression analysis for OCD disorder as dependent variable in different groups of patients with epilepsy

As can be seen from Tab 3 in the final predictive score of OCD the several personality variables are important. These variables included mostly such features, as Alexithymia, Extraversion, and Neuroticism. Noteworthy, Alexithymia has been included into regression equation with positive loadings in left-handers and in group of left-handers with right focus activity. It implies the alexithymia to be a risk factor for OCD development strictly in left-handed TLE patients and particularly in left-handed persons with right focus activity. On the other hand, Alexithymia has been included into regression equation with negative loading in group of right-handed patients with left-sided focus epileptic activity. It implies that in such category of patients Alexithymia reduces the final OCD score and by that exerts protective action against its development.

Noteworthy, the Extraversion and Neuroticism scores both have been included into regression within same several combination of handedness and focus laterality, although with opposite loadings. Thus, Extroversion always had negative loading and by that reduced the final OCD score. It concerned as all patients, as especially patients with right focus activity and the right-handed patients and persons with combination of right-handedness and right focus activity. In other words the high Extraversion exerts protective effect on OCD development, while low level of Extraversion (Introversion) determines OCD and right-handed persons with right focus activity are especially vulnerable to it.

Neuroticism, on the contrary had been included into final regression equation with positive loadings and by that determined the OCD development in the same categories of patients. In other words, Extroversion and Neuroticism have an opposite influence on OCD score, and Neuroticism is thought to be a risk factor for OCD development.

On the final stage comparison of mutual variances between dependent and independent variables in regression analysis has been performed (Tab 3). The comparison revealed that maximal values of variances were observed in regression performed for group of patients with right focus activity and especially for group of right-handers with right-sided focus (R2=0,74) and for person with combination of left-handedness and left focus activity (R2=0,78).

Nevertheless, the role of maximal variance value in the last category of patients (left-handers with left focus activity) shouldn't not been exaggerated, since no statistically significant loading for any personality construct has been observed.

Conversely, the minimal values of explained variances were obtained for whole group of patients where handedness and focus laterality have not been taken into account

5. Discussion

The current study is rather the second one, in which the premorbid personality features were evaluated as risk factors in combination with certain neurobiological variables in patients with TLE. Earlier we have performed the similar work on relationships between premorbid personality traits and development of depression and anxiety states in temporal lobe epilepsy [29]. The principal limitations of the current study concern the small size of LHRF and LHLF groups, that couldn't reveal always statistically significant correlations and discrepancies between compared groups. Nevertheless, obtained data have shown that based on certain so-called premorbid variables rated subjectively by patients themselves the prediction of OCD in TLE patients is quite possible.

Principally, the identification and diagnostics of OCD pathology have been performed by psychiatrists using objective scales, and those physicians were blind to data on patient personality traits. Such approach has been used in order to avoid any bias in the qualification of patient personality structure.

Another reason for the criticism is statement that traits used in study depict not premorbid period in TLE patients at all, but personality changed due epilepsy course itself. Here should be stressed that similar approach has been already widely used in numerous studies on different contingent of patients including organic brain disorders and their authors believe that MPT constructs can assess strictly premorbid personality [15-18]. Moreover, personality changes in TLE patients are characterized as a rule by other kind of features, and viscosity is the prominent trait among them and it isn't included into either construct of MPT. In this context MPT doesn't depict the structure of epileptic personality at all, although Rigidity construct can be seen as similar but not identical to epileptic personality changes.

The principal results of present study have shown, that neither seizure semiotics, nor focus lateralization (assessed separately from personality features) couldn't discriminate epileptic patients in terms of risk OCD development. However, such prediction becomes probable, if take into account the personality traits.

The third reason for critic of current study may concern the fact that here has not been used Yale-Brown scale. Here should be stressed that OCD construct included in SCL-90 questionnaire describes thoroughly obsessive-compulsive symptoms and may be adequately enough used for the assessment of such signs severity along with other psychopathological constructs [25, 26].

The main results of the current study have shown that premorbid personality features have multifactorial influence on OCD development in patients with TLE. Thus, obsessionality in patients with TLE may evolve similarly from such different personality traits as alexithymia, low extraversion score (introversion) and neuroticism and esoteric tendencies. Nonetheless, the certain premorbid personality trait becomes relevant for OCD development strictly on condition that some definite neurobiological variables, such as handedness, focus lateralization and their combination in TLE patients should also be taken into account. Moreover, the certain interaction among concrete personality characteristics with definite neurobiological variables indeed exists.

Thus, Alexithymia may become risk factor for OCD appearance strictly in left-handed patients and in left-handed persons with right focus activity. Quite the contrary, Alexithymia may play protective role against OCD development in right-handed patients with left focus activity. In other words, quite mirror image role of Alexithymia for OCD origin in epilepsy depends on interaction between handedness and focus lateralization.

Neuroticism and Extroversion, on the other hand, were risk factors for OCD in cases of right-handed patients with right-focus activity, although Neuroticism had positive loading for OCD development in right-handed patients with the left focus activity too. Obviously, the right-handedness seems to be more important for Neuroticism construct in terms of its role in OCD triggering than focus laterality. It implies that Neuroticism as personality feature may be maximally expressed in the right-handed, but not in the left-handed persons, i.e. seems to be the prerogative of persons with normal motor lateralization.

Obtained results have confirmed data of our previous work, that Alexithymic and Neurotic traits are thought to depict the different types of personality that rather exclude each other. In this context the appearance of similar OCD disorders in TLE patients with such different personality structures seems to be unexpected and rather contradictory and should be properly explained. Unfortunately, only speculative explanation may be proposed in this context. Obviously, the OCD and affective and anxiety disorders seem to be the adaptive forms of reaction on environmental and internal factors, and could appear in evolution of animals and *Homo sapiens*. As adaptive forms of behavior these disorders are required in persons with different types of neurobiological mechanisms, including motor lateralization and focus laterality in TLE.

Contrary to our initial expectations we couldn't find any significant effect of construct "Rigidity" on OCD in patients with TLE, although the repetition of obsessions seems to be an attribute of rigidity. Principally, in our previous work we couldn't find any relationship between Rigidity and depression development in patients with epilepsy [29]. It implies that "Status melancholicus" concept is not rather relevant enough for prediction of affective disorders and OCD development, and mentioned psychiatric co-morbidity have different personality predisposition traits that become trigger risk factors in TLE. Obviously, that alexithymia, extraversion, neuroticism and esoteric tendencies are more relevant for depression, anxiety and OCD in persons with TLE, than Rigidity.

Based on the data obtained in the current and previous work [29] conclusion can be made, that premorbid personality profile of affective, anxiety disorders and OCD is quite similar and includes high Neuroticism level and low Extraversion (high Introversion) level in the right-handed patients with right focus activity. Besides, the high level of Alexithymia seems to be *Conditio sine qua non* for the OCD and depression and anxiety in left-handed persons. Obviously that Neuroticism combined with low level of Extraversion (Introversion), on one side, and Alexithymia, on the other side, may cause the similar psychopathological disorders such as OCD and depression and anxiety states in patients with different neurobiological predisposition. It implies that Neuroticism, Extraversion and Alexithymia represent the universal personality constructs for development as affective, as OCD. Further studies are required to elucidate the pathogenic mechanisms of co-morbid psychopathology in person with epilepsy.

Author details

Vladimir V. Kalinin, Anna A. Zemlyanaya, Elena V. Zheleznova and Lyudmila V. Sokolova

Department of Brain Organic Disorders and Epilepsy of Moscow research Institute of Psychiatry of Ministry of Health and Social Development, Russia

References

[1] Perini G.I., Tosin C., Carraro C., et al. Interictal mood and personality disorders in temporal lobe epilepsy and juvenile myoclonic epilepsy. J Neurol Neurosurg Psychiatry 1996; 61 (6): 601 – 605.

[2] Hermann B.P., Seidenberg M., Woodard A. et al. Psychiatric comorbidity in chronic epilepsy: identification, consequence, and treatment of major depression. Epilepsia 2000; 41: S31-41.

[3] Mendez M.F., Cummings J.L., Benson D.F. Depression in epilepsy. Significance and phenomenology. Arch Neurology 1986; 43: 766-770.

[4] Mendez M.F., Taylor J.L., Doss R.C., Salguero P. Depression in secondary epilepsy: relation to lesion laterality. J.Neurol. Neurosurg. Psychiatry 1994; 57: 232-233.

[5] Kanner A., Nieto J. Depressive disorders in epilepsy. Neurology 1999; 53 (Suppl.2): S26 – S32.

[6] Kanner A. Depression in neurological disorders. Lundbeck Institute, Cambridge Medical Communication Ltd, 2005 – 161pp.

[7] Monaco F., Cavanna A., Magi E. Et al. Obsessionality, obsessive-compulsive disorder, and temporal lobe epilepsy. Epilepsy & Behavior 2005; 7(3): 491-496.

[8] Bear D.M., Fedio P. Quantitative analysis of interictal behavior in temporal lobe epilepsy. Arch Neurology 1977; 34: 454-467.

[9] Rodin E., Schmaltz S. The Bear-Fedio personality inventory and temporal lobe epilepsy. Neurology 1984; 34: 591-596.

[10] Gupta A.K., Ettinger A.B., Weisbrot D.M. Psychiatric comorbidity in epilepsy. In: Ettinger A.B.,Devinsky O.(Eds) Managing epilepsy and co-existing disorders. Butterworth-Heinemann, Boston, 2002: 343-387.

[11] Tuke D.H. Imperative ideas. Brain, 1894, 17:179-197.

[12] Berrios G.E. The history of mental symptoms. Descriptive psychopathology since the nineteenth century. Cambridge University Press, 2002 – 565 pp.

[13] Kalinin V.V., Zemlyanaya A.A., Krylov O.Y., Zheleznova E.V. Handedness, alexithymia and focus laterality as risk factors for psychiatric co-morbidity in patients with epilepsy. Epilepsy &Behavior, 2010.

[14] Tellenbach H. Melancholie. Problemgeschichte, Endogenität, Typologie, Pathogenese, Klinik, 3. Auflage, Springer, Berlin, 1971. (in German).

[15] Zerssen D. von. Objektivirende Untersuchungen zur prämorbiden Persönlichkeit endogenen Depressiver (Methodik und vorläufig Ergebnisse). In: Hippius H., Selbach

H. (eds.). Das depressive Syndrom. Urban und Schwarzenberg, München, 1969: 183-205 (in German).

[16] Zerssen D. von. Personality and affective disorders. In: Paykel E.S. (eds.) Handbook of Affective Disorders. Churchill Livingstone, Edinburgh, 1982: 212-228.

[17] Möller H-J. Zur Bedeutung und methodischen Problematik der psychiatrischen Persönlichkeitsforschung: der Typus melancholicus und andere Konzepte zur prämorbiden Persönlichkeit von Patienten mit affektiven Psychosen. In: Marneros A., Philipp M.(eds.). Persönlichkeit und Psychische Erkrankung. Springer-Verlag, Berlin, 1992; 45-65 (in German).

[18] Marneros A., Deister A., Rohde A. Prämorbide und postmorbide Persönlichkeitsmerkmale bei Patienten mit idiopathischen Psychosen. In: Marneros A., Philipp M. (eds.). Persönlichkeit und Psychische Erkrankung. Springer-Verlag, Berlin, 1992; 87-100 (in German).

[19] Furukawa T., Nakanishi M., Hamanaka T. Typus melancholicus is not the premorbid personality trait of unipolar (endogenous) depression. Psychiatry and Clinical Neurosciences 1997, 51: 197-202.

[20] Zerssen D. von, Pfister H., Koeller D-M. The Munich Personality Test (MPT): a short questionnaire for self-rating and relatives' rating of personality traits: Formal properties and clinical potential. Eur Arch Psychiatry Neurol Sci, 1988, 238: 73-93.

[21] Eysenck H.J. & Eysenck S.B.G. Manual of the Eysenck Personality Inventory. London, Hodder & Stoughton, 1975.

[22] Kretschmer E. Körperbau und Charakter, 26 Auflage, Springer-Verlag, Berlin, 1977 (In German).

[23] Bagby R.M., Taylor G.J., Ryan D. Toronto Alexithymia Scale: Relationship with personality and psychopathology measures. Psychother Psychosom., 1986, Vol.45, P. 207-215.

[24] Bagby R., Taylor G., Parker J. Cross-validation of the factor structures of the Toronto alexithymia scale. J. Psychosom. Res., 1990, Vol. 34, P.47 – 51.

[25] Derogatis L.R., Lipman R., Covi L. SCL-90: an outpatient psychiatric rating scale: preliminary report. Psychopharmacol Bull, 1973; 9: 13-28.

[26] Derogatis L.R., Cleary P.A. Confirmation of the dimensional structure of SCL-90: a study of construct validation. J Clin Psychol 1977; 33: 981-989.

[27] Annett, M. (1970). A classification of hand preference by association analysis. British Journal of Psychology, 61, 303-321.

[28] Feinstein A.R. Principles of medical statistics. Boca Raton, Chapman&Hall/CRC, 2002-701 P.

[29] Kalinin V.V., Zemlyanaya A.A., Zheleznova E.V,, Krylov O.E Premorbid Personality Traits, Focus Lateralization and Handedness as Risk Factors for Co-Morbid Affective and Anxiety Disorders in Temporal Lobe Epilepsy // Horizons in Neuroscience Research, Volume 7/Eds. Andres Costa and Eugenio Villalba, Nova Publishers, 2012- pp. 175-190.